# The Japanese numbers game

T0174985

# The Nissan Institute/Routledge Japanese Studies Series

*Editorial Board:*

J. A. A. Stockwin, Nissan Professor of Modern Japanese Studies, University of Oxford and Director, Nissan Institute of Japanese Studies

Teigo Yoshida, formerly Professor of the University of Tokyo, and now Professor, The University of the Sacred Heart, Tokyo

Frank Langdon, Professor, Institute of International Relations, University of British Columbia, Canada

Alan Rix, Professor of Japanese, The University of Queensland

Junji Banno, Professor, Institute of Social Science, University of Tokyo

*Other titles in the series:*

*The Myth of Japanese Uniqueness*, Peter Dale
*The Emperor's Adviser: Saionji Kinmochi and pre-war Japanese Politics*, Lesley Connors
*Understanding Japanese Society*, Joy Hendry
*Japanese Religions*, Brian Bocking
*Japan in World Politics*, Reinhard Drifte
*A History of Japanese Economic Thought*, Tessa Morris-Suzuki
*The Establishment of Constitutional Government in Japan*, Junji Banno, translated by J. A. A. Stockwin
*Japan's First Parliaments 1890–1910*, R. H. P. Mason, Andrew Fraser, and Philip Mitchell
*Industrial Relations in Japan: the peripheral workforce*, Norma Chalmers
*Banking Policy in Japan: American attempts at reform during the occupation*, William Minoru Tsutsui
*Educational Reform in Contemporary Japan*, Leonard Schoppa
*How the Japanese Learn to Work*, Ronald Dore and Mari Sako
*Militarization in Contemporary Japan*, Glenn Hook
*Japanese Economic Development: theory and practice*, Penelope Francks
*Japan and Protection*, Javed Maswood
*Japan's Nuclear Development*, Michael Donnelly
*The Soil, by Nagatsuka Takashi: a portrait of rural life in Meiji Japan*, translated and introduced by Ann Waswo
*Biotechnology in Japan*, Malcolm Brock
*Britain's Educational Reform: a comparison with Japan*, Michael Howarth
*Language and the Modern State: the reform of written Japanese*, Nanette Twine
*Industrial Harmony in Modern Japan: the invention of a tradition*, W. Dean Kinzley
*Japanese Science Fiction: a view of a changing society*, Robert Matthew

# The Japanese numbers game

The use and understanding of numbers in modern Japan

Thomas Crump

Routledge
Taylor & Francis Group

LONDON AND NEW YORK

First published 1992
by Routledge
2 Park Square, Milton Park, Abingdon, Oxon, OX14 4RN

Simultaneously published in the USA and Canada
by Routledge
605 Third Avenue, New York, NY 10017

*Routledge is an imprint of the Taylor & Francis Group, an
informa business*

© 1992 Thomas Crump

Typeset in Times Roman by Leaper & Gard Ltd, Bristol

All rights reserved. No part of this book may be reprinted or
reproduced or utilized in any form or by any electronic,
mechanical, or other means, now known or hereafter
invented, including photocopying and recording, or in any
information storage or retrieval system, without permission in
writing from the publishers.

Notice:
Product or corporate names may be trademarks or registered
trademarks and are used only for identification and explanation
without intent to infringe.

*British Library Cataloguing in Publication Data*
Crump, Thomas
    The Japanese numbers game: the use and understanding of
    numbers in modern Japan.
    1. Japan. Numbers. Sociology
    I. Title    II. Nissan Institute of Japanese Studies
    306.45

    ISBN 0–415–05609–8

*Library of Congress Cataloging in Publication Data*
Crump, Thomas.
    The Japanese numbers game: the use and understanding of numbers
    in modern Japan / Thomas Crump.
        p.    cm. — (The Nissan Institute/Routledge Japanese Studies
    series)
    Includes bibliographical references and index.
    ISBN 0–415–05609–8
    1. Folklore—Japan.    2. Numerals—Folklore.    3. Symbolism of
    numbers—Japan.    4. Japan—Civilization—1945-    5. Japan—Social
    life and customs.    I. Title.    II. Series.
    GR340.C78    1992
    398′.356′0952—dc20                                            91–10047
                                                                  CIP

ISBN 13: 978-0-415-05609-0 (hbk)

# Contents

| | | |
|---|---|---|
| *List of figures* | | vi |
| *List of tables* | | vii |
| *General editor's preface* | | viii |
| *Preface* | | x |
| 1 | **The numerical paradox** | 1 |
| 2 | **Numbers in the written and spoken language** | 14 |
| 3 | **Alternative number systems** | 33 |
| 4 | **The culture of numbers** | 46 |
| 5 | **What's in a Japanese name?** | 63 |
| 6 | **Fortune-telling** | 76 |
| 7 | **Time** | 96 |
| 8 | **The spatial world of numbers** | 114 |
| 9 | **The Japanese abacus** | 126 |
| 10 | **Games – ancient and modern** | 139 |
| 11 | **The ecology of numbers – past and present** | 155 |
| | *Notes* | 163 |
| | *References* | 196 |
| | *Index* | 202 |

# Figures

| | | |
|---|---|---|
| 3.1 | Simple and complex mandala | 38 |
| 3.2 | Gogyō | 40 |
| 3.3 | Kanshi | 42 |
| 4.1 | Why 10 = 10,000 | 55 |
| 5.1 | Seimeigaku | 69 |
| 5.2 | Gogyō equivalents of the kana syllabary | 71 |
| 6.1 | Omikuji temple scene | 78 |
| 6.2 | Omikuji written advice | 79 |
| 6.3 | The basic hōiban | 86 |
| 6.4 | The directional magic square | 87 |
| 6.5 | The order of numbers in the nine hōiban in the kyūsei cycle | 88 |
| 6.6 | Kyūsei cycles of good and ill fortune | 90 |
| 7.1 | Yakudoshi | 103 |
| 7.2 | The ritual cycle of life and death | 105 |
| 7.3 | The production cycle in the perspective of life and death | 108 |
| 8.1 | Japanese floor-plans | 123 |
| 9.1 | The Japanese abacus | 128 |
| 9.2 | Abacus multiplication | 129 |
| 10.1 | Sumo: the Kanroku scale | 152 |

# Tables

| | | |
|---|---|---|
| 2.1 | Chinese numerals | 26 |
| 3.1 | Trigrams | 37 |
| 3.2 | Jikkan jūnishi | 44 |
| 6.1 | Nenrei hayamihyō | 82 |
| 6.2 | Nenrei hayamihyō (English version) | 83 |
| 6.3 | The nine kyū | 84 |
| 6.4 | Affinity between man and woman | 85 |
| 6.5 | The lines of trigrams and hexagrams | 93 |

# General editor's preface

Now that we are in the 1990s, most would agree that Japan is a nation of absolutely first-rate significance. The successes of the Japanese economy and the resourcefulness of her people have long been appreciated abroad. More and more people are also becoming aware of the increasing impact of Japan on the outside world in a widening variety of fields of endeavour. This tends to produce painful adjustment and uncomfortable reactions. It also leads to the formation of stereotypes and arguments based on outdated or ill-informed ideas.

The Nissan Institute/Routledge Japanese Studies Series seeks to foster an informed and balanced – but not uncritical – understanding of Japan. One aim of the series is to show the depth and variety of Japanese institutions, practices and ideas. Another is, by using comparison, to see what lessons, positive and negative, can be drawn for other countries. There are many aspects of Japan which are little known outside that country but which deserve to be better understood.

The Japanese popular use and understanding of numbers is something which is hardly known at all outside Japan. Dr Thomas Crump, writing from the perspective of social anthropology, introduces the reader in fascinating detail to the world of numbers in which fortune-telling, the abacus, games involving numbers as well as curious numerical names (of both people and places) illustrate a popular obsession with systems of counting, calculation and forecasting. The cultural roots of attitudes towards numbers are meticulously explored, and suggestions are made about the contemporary implications of a culture in which mechanical numeracy (and number obsession) is general but the highest levels of academic mathematics still fall short of world standards. Dr Crump takes the reader through intriguing byways which suggest that many Japanese people truly treat numbers as their friends.

J. A. A. Stockwin
Director, Nissan Institute of Japanese Studies
University of Oxford

# Preface

Almost on my first day in Japan, in the summer of 1980, I was struck by the compulsive interest of the Japanese in numbers. On a visit to the Imperial Palace Gardens, two young men politely asked me whether they could talk to me in English. Within a very short time the conversation had turned to the question of the difference in use between the words *shi* and *yon* for the number 4 and I heard for the first time that 'four men' must be '*yon*in', because *shi* also meant 'death'. The conversation then turned to the *gojūon*, the fifty sounds comprised in the two Japanese syllabaries, and it began to dawn upon me that the Japanese had no difficulty in talking about numbers. In the following years, during which I was to visit Japan five times, sometimes for periods of several months, this impression was confirmed time and time again.

The last few days of that first visit I spent in Kyoto. From the tourist literature supplied to me it became clear that I ought to visit the temple of Sanjūsangendō. This was within walking distance of the small *ryokan* where I was staying. As I followed the map, I suddenly came across a signpost, with the name 三十三間堂, and this triggered the 'Aha!' reaction necessary to guide my future research. The Chinese characters, or *kanji*, used to write almost all names in Japanese make no distinction between numbers and other components of a word. The number 33 (*san-jū-san*) is embedded in 'Sanjūsangendō', denoting the number of bays (*gendō*) which give the temple its name. Sanjūsangendō is perhaps an extreme case of number symbolism by reason of the immense amount incorporated into the temple and its contents (which I describe in detail in chapter 8), but it is by no means an isolated instance of this phenomenon.

Numbers occur everywhere in Japanese popular culture. In a single book it has been quite impossible to cover the whole ground. At the same time much of the material will be familiar to readers who already know Japan. I am confident, however, that even the most blasé reader will encounter the odd shaft of light shone upon some hitherto unknown aspect of Japanese culture. At first sight the available material seems so varied as to defeat any attempt at systematic presentation. Some sort of taxonomy of numbers is possible, however, and this is indicated by the contents list.

A number of principles have guided me in selecting material. First and foremost, the use and understanding of numbers, in the cases I present, are both popular and contemporary. Nothing is said about esoteric cults confined to some particular class or region. On the other hand, 'popular' does not necessarily mean 'universal'. To give one example, although the use of the abacus is widespread throughout Japan, one can easily find Japanese – typically the young *sarariman* – who have forgotten all they ever learnt. At the same time it is ubiquitous in the millions of small shops and market stalls so characteristic of Japanese retailing, where the level of proficiency is often breathtaking.

'Contemporary', also, does not exclude ancient origins. Sanjūsangendō is several hundreds of years old, but although the part it plays in the life of Kyoto has undoubtedly changed, it is still very much part of the local culture – as is obvious to any visitor. On the other hand, some traditional numerical institutions are not dealt with, simply because they have been buried by popular equivalents from the western world. This is one reason why I say nothing about *wagaku*, or autochthonous Japanese music. Although this music is still played, one would have to take a very optimistic view of Japanese culture to describe it as 'popular'. In this case also, the complexities of Japanese musical notation inhibited me from any serious discussion.

*Wagaku* is but one instance of autochthonous cultural institutions which can be identified by the prefix 'wa' 和, which to the Japanese generally connotes 'peace' or 'harmony'. In the context of numbers, there is also *wasan*, a sort of traditional Japanese arithmetic. This I have also left out of account, for it is neither popular nor contemporary, even though there are quite a number of simple books, intended for the general public, which describe it. Arithmetic, as a school subject, is not presented in any way as part of Japanese culture, and the Ministry of Education banished *wasan* from the syllabus more than a hundred years ago.

A further principle has restricted the choice of subject matter. Everything I discuss is in some way distinctively and characteristically Japanese. This has largely excluded any discussion of numbers in practical technological uses or in cultural contexts derived from the west. The principle allows discussion of the abacus or sumō, but not of the computer or baseball, even though computer freaks and baseball fans may be just as numerous in Japan as in the United States. The point I am continually trying to make is simply that the intensive use of numbers in Japan, in any number of different and often surprising contexts, is not only highly original but also provides a key to understanding the Japanese mentality.

In a book written mainly for western readers, this last point deserves one particular comment. In the west one can still find people who, while considering themselves as members of some cultural elite, affect a disdain for anything connected with the exact sciences. Commerce is quite beyond the pale, even though in many cases it was commercial success in the ancestral line which gave the elite its privileged position. Attitudes of this kind rule out any aesthetic appreciation of numbers.

In pre-modern Japan such attitudes were characteristic of the samurai, who confined their mathematical interests to *wasan*. My research in modern Japan revealed hardly any trace of them. There is no disdain for numbers. As for myself, the last thing I want to achieve is that readers should use the material in this book to confirm some feeling of cultural superiority. Even such aberrations as the use of the game of *janken* (which I describe in chapter 10) for the selection of *kamikaze* pilots have their parallels in the western world.

My research was made particularly agreeable by the way in which Japanese friends and colleagues welcomed my interest in the use and understanding of numbers in modern Japan. The subject brought me into contact with an astonishing range of people, in almost every part of Japan. Market stalls, fishing boats, youth hostels, abacus schools, computer laboratories, school playgrounds, family living-rooms, the slopes of any number of mountains (including Fuji), shrines and monasteries, and even cemeteries, combined to set the scene for my research. In all these differebt places I found more than sufficient help and interest to keep me going.

At every stage my Japanese colleagues willingly answered my questions, suggested lines for further enquiry, and arranged for me to meet those who could help me on particular aspects of number use. In particular I should like to thank the following for their constant help and advice: Prof. M. Akasaka of the University of Toyama, Prof. H. Araki and Prof. Y. Yano of the Education University of Kyoto, Dr H. Nakamaki and Dr E. Kuroda of the National Museum of Ethnology in Osaka, Prof. T. Hatta of the Osaka Education University, Prof. T. Hayashi of the Doshisha University, Kyoto, Prof.

H. Saito of the University of Nagoya and Prof. T. Yoneyama of the University of Kyoto. Outside Japan I would like to thank Mr John Breen of the School of Oriental and African Studies in London, Prof. Sepp Linhart of the University of Vienna, Prof. Emiko Ohnuki-Tierney of the University of Wisconsin, and Dr Ian Reader of the University of Stirling, all of whom gave me invaluable help on matters about which they were much better informed than I. Just before this book went to press, I had the opportunity of continuing the field-work on which it is based among the Japanese community in Hawaii. Unfortunately this was too late to work the results into the book, although I have been able to add one or two footnotes. I am still most grateful fo Professor Willa Tanabe of the Center for Japanese Studies at the University of Hawaii at Manoa, and also many colleagues at the Center, for providing me with the most generous academic hospitality during my stay. I am equally grateful to Professor Franklin Odo and Professor Dennis Ogawa, whose knowledge of the culture developed by Americans of Japanese ancestry is quite unequalled, for providing me with the benefit of their experience. My gratitude extends to many not mentioned above, particularly in all the different parts of Japan and Hawaii I visited in the course of field-work.

The research on which this book is based would never have been possible without the support of the Faculty of Political, Social and Cultural Sciences of the University of Amsterdam, which also provided much of the necessary funding. My colleagues in the Faculty Group, Cultural Anthropology, always enabled me to find the time needed for my visits to Japan; I am grateful also to all those students who accepted the necessary and often inconvenient changes in the time-table. Above all, my thanks are due to the Japan Foundation, which appointed me a fellow for the year 1987, enabling me to spend a much longer time on research in Japan than would otherwise have been possible.

Much of the theoretical material is to be found in my book *The Anthropology of Numbers*, which was published earlier this year. I am most grateful to Cambridge University Press for their permission to use and reproduce this material in the present book.

Finally I should like to thank my wife, Carolien, and my children, Maarten and Laurien, for putting up with me when my mind was often far away in the world of Japanese numbers. They may find it difficult to believe, but their forbearance is greatly appreciated.

# 1 The numerical paradox

## THE PROBLEM STATED

Any number of questions arise from the Japanese preoccupation with numbers. Intuitively numbers seem so much a part of all human experience, that it is difficult to see precisely what scope there is for understanding, or using them, in some special way.[1] True, certain specialised numerical institutions, such as certain games, may be found to exist only in one particular region – which for this book will be Japan – but even then the actual numerical base can always be expressed in a form which is quite general and not tied to any particular culture.[2] The symbols used tend to be completely arbitrary, whether they take the form of spoken words or written signs, which need not necessarily correspond to one particular word. Viewed objectively, this is true of the numerical symbols used by the Japanese, although this is not how the Japanese themselves would see things. It is not so much that certain symbols, such as the Chinese character, or *kanji*,[3] for the number 3, 三, indicate their meaning by being a member of the class defined by this number, in the present case by being comprised of three horizontal strokes, but that many other *kanji* numerals, without this property, still have connotations which range far beyond the numbers they represent. Instances will constantly occur throughout this book.

This is the heart of the paradox inherent in the use and understanding of numbers in any culture. In one direction, numbers can always be reduced to a complete abstraction, whose properties give no indication of how they might be used in any practical or cultural context. This is the direction followed by pure mathematics, whose practitioners tend to eschew all practical applications.[4] In the other direction, numbers are important not for their inherent properties, such as are the basis of any sort of arithmetic,[5] but for what they connote, and this, far from being general, is highly particular. To give but one example, already noted in the preface, one of the two words for 4 in Japanese, *shi*, also means 'death', which is taken by many Japanese to explain why the word for 'four men', *yonin*, is based on the alternative *yon*. Japanese culture is extraordinarily rich and creative in such connotations. More than this, it extends the process to identifying certain numbers on the basis of their inherent arithmetical properties and then seeing them as having a special meaning. A typical instance of this, to be looked at in detail later, is to regard numbers, such as 33, which exceed by 1 a power of 2 (so that $33 = 2^5 + 1$), as specially significant.

All this is a part of what is now, fashionably, called 'cognition'. Its basis is the knowledge of the world drawn on to construct perceptions of it (Keesing 1981: 97); this is what Gregory (1969) has described as an 'internal model of reality'. Cognition is for the individual what culture is for society, but, inescapably, culture provides the individual with the means to build his own models. These means, particularly when it comes to numbers, are symbolic. In the result

> when seen as a set of symbolic devices for controlling behaviour ... culture provides the link between what men are intrinsically capable of becoming and what they actually, one by one, in fact become. Becoming human is becoming individual, and we become individual under the guidance of cultural patterns, historically created systems of meaning in terms of which we give form, order, point and direction to our lives. And the cultural patterns involved are not general but specific. (Geertz 1973: 52)

Turning to Durkheim, although Japan is far removed from the *lower* societies upon which he based his study of religious forms, one's intuitive reaction is still likely to be that

> the slighter development of individuality, the small extension of the group, the homogeneity of external circumstances, all contribute to reducing [cultural] differences and variations to a minimum. The group has an intellectual and moral conformity of which we find but rare examples in [other] advanced societies. Everything is common to all. Movements are stereotyped; everybody performs the same ones in the same circumstances, and this conformity of conduct only translates the conformity of thought. Every mind being drawn into the same eddy, the individual type nearly confounds itself with that of race. And while all is uniform, all is simple and well. (Durkheim 1915: 5-6)

If, in relation to Japan, this passage from Durkheim represents both the impression the land makes upon the casual visitor from overseas and the ideal which many Japanese would like to see realised, it is still largely a parody of the true situation. Little is simple about the way Japanese use and understand numbers. Many of the contexts and forms of such use are outside the experience of millions of Japanese, whether by accident or design. Where all the members of one family may be named according to the principles of *seimeigaku* examined in chapter 5, in another, these may be completely disregarded. Even within the family differences will occur: a businessman, who has forgotten everything he ever knew about the abacus, may have a wife who uses it expertly for managing the family budget. But then once again citing Durkheim (1915: 2–3):

> The most ... fantastic rites and the strangest myths translate some human need, some aspect of life, either individual or social. The reasons with which the faithful justify them may be, and generally are, erroneous; but the true reasons do not cease to exist, and it is the duty of science to discover them.[6]

Science, if it is to discover the reasons which explain all the different uses of numbers in Japan, must look in two directions. The one is nature, the other is cosmos, and it is in coming to terms with nature and cosmos, over a period of some 2,000 years, that the Japanese have developed all their different numerical institutions. What then are nature and cosmos, and what then is their relation with numbers?

Nature and cosmos have in common that they are that part of the environment which are given in the experience of any community. They are manifest in immutable laws, which define the scope of all human initiative. Man cannot escape from the laws of nature, nor from the cosmic order. At the same time such law and order are a projection of purely cultural forms, although the culture from which they derive makes certain that their origin is suppressed. The point is simple enough: culture continually improves upon nature, and justifies the rules necessary for achieving and sustaining any improvement by an appeal to the abstract order of the cosmos. The process can be illustrated by the case of wet rice cultivation in Japan, which at the same time may be used to point out the essential difference between nature and cosmos.

The cultivation of rice, like any other form of agriculture, cannot escape from the inherent properties of the plant upon which it is based. At the same time, the technology developed within any culture can ensure that these properties are exploited so that yields far exceed what nature, left to itself, could ever produce.[7] Nowhere has this process gone further than it has with the cultivation of rice on irrigated terraces, as practised in Japan for well over a thousand years.[8] The rice terraces reflect not only the vast investment in human capital required for their construction and maintenance, but also provide for the sustenance and livelihood of the population which this represents. Wild rice, if it ever existed in Japan, could never have provided the basic means of subsistence.[9]

The success of wet rice cultivation was tied to the use of numbers in two quite different ways. In the first place, the terrace system of cultivation, allowing for no economies of scale (Bray 1986: 15), always depended upon the co-ordination of the work of a large number of small family farms. The economic problem of regulating land and labour was essentially numerical, the more so when the heavy tax burden resting on the village fell to be shared by the different households comprising it.[10] In the second place, the cultivation of rice was still dependent upon the annual cycle of the seasons, as recorded in the calendar, which was itself determined by the cosmic order. This in turn gave rise to considerable numerical problems (which I discuss further in chapter 7, 'Time'), particularly in relating the length of the month (as determined by the moon) to that of the year (as determined by the sun).

One faces here a critical distinction between nature and cosmos. Although the course of nature up to a certain point can be directed and contained (which fact underlies all successful cultivation), beyond this point it is unpredictable and uncontrollable. The Japanese *sansai*, or three natural calamities of fire, storm and flood, occur capriciously, leaving behind devastation which no advance planning can counteract. Events of this kind can be accounted for *post hoc*, sometimes by being related to such cosmic events as eclipses, which occur outside the normal calendar.[11]

The cosmic order is, in principle, predictable (for otherwise it would be *disorder*) and not so much uncontrollable as simply beyond control. Knowledge of the cosmic order is a matter of observation and deduction, requiring the keeping of records and arithmetical calculation for the purpose of interpreting them. Both of these operations are essentially numerical: the first requires the linguistic base described in chapter 2; the second, an efficient instrumental means such as is provided by the abacus described in chapter 9.

The object is not to master the cosmic order, which is impossible, but to understand and explain it. The question is: what sort of explanations are acceptable in the local culture? In answering this question one comes to a parting of the ways between east and west. Modern western man accepts a disjunctive view of the universe, which sees no necessity for the observations of astronomy to regulate day-to-day life, save in so far as the phenomena observed control natural events, such as tides and seasons. As Needham (1969: 26) puts it, 'from the beginning of their thought history, Europeans have passed continually from one extreme world outlook to another, rarely finding any synthesis.... Theological spiritualism and mechanical materialism maintained perpetual war.'[12] The result is that cosmology is essentially demythologised; its basis is logical, and it is a part of mathematics. This is the world of modern, western astronomy, in which mathematical notation ignores, so far as possible, the traditional configuration of the heavens.[13]

The traditional world of Japanese numbers has another interpretation of cosmology. The order, reflected in the universe, is a conscious order, a manifestation of some ideal plan, so that the phenomena observed can be interpreted as signals requiring an appropriate response from the observers. Solstices and equinoxes are imperatives in a moral order, expressed in the calendar, in a way explained in detail in chapter 7.

## THE TYPOLOGY OF NUMBERS

The discovery of the true nature of numbers is a logical problem inherent in any culture in which numbers occur.[14] Few, if any, cultures have ever achieved an adequate solution, for the problem itself is extremely difficult to formulate.[15] To begin with the most elementary level of cognition – which may be taken to be that of the child as it first becomes familiar with numbers – numbers are intuitively associated with counting. But counting, since it assumes the existence of numbers, cannot be any part of their logical basis. Judged culturally, the essential intuition is that numbers are *there*, whether or not counting, or any other process, is necessary to make them part of everyday cognition. Numbers in this elementary sense are generally known as *natural numbers*, so that instances of numbers are no more than events necessary to bring such numbers within human experience.

To speak simply of 'numbers' or 'natural numbers' is an oversimplification. Any analysis of contexts in which numbers occur will disclose two logical types, ordinal and cardinal, each with their own distinctive lexical forms (considered further in chapter 2).[16] Both types presuppose the existence of *sets* containing more than one member, and capable of being submitted to some sort of intellectual process, which will determine in the case of every set some number characteristic of it. The process is inevitably linguistic,[17] and so in practice will involve counting, at least implicitly.

Order is a property of the members of any set in which there is a recognised relation of precedence and succession. Such a relationship imposes a unique order upon all the members of the set. Theoretically there may be a problem in establishing the relationship, but in practice circumstance tends to define the principle to be applied. Time is the classic case, since in any set based upon points in time (which may, for instance, be the dates of birth of the members of a family) the concept of time automatically determines the ordering of these points. In particular the cosmic order defines a number of such sets, familiar to us as the succession of days, months or years. The principle so established can be extended to create, within any culture, such constructs as a 'week' or a 'decade',[18] which have no cosmic basis.[19]

In practical terms, putting the members of a set in a given order implies that their number is not so great as to frustrate this process. Where this number is small, as it tends to be in traditional cultures, there is hardly any problem, and the ordering process simply equates the members of the class with the succession of natural numbers, 1, 2, 3, up to the limit defined by the total number of members. In almost every case, where these numbers are explicitly retained, they have a distinctive lexical form, such as the English, *first, second, third* and so on. The fact that an ordered class must be closed does not prevent the events ordered by it repeating themselves indefinitely. In such a case, the standard form of ordinals is often replaced by special forms, such as the days of the week or the notes of the octave. As I show in later chapters of this book, Japan provides an exceptionally wide range of instances of this practice; of these, some, such as the months of the year (but not the days of the week) are explicit in their use of the standard numbers, where others, such as the positions taken in the game of *janken* (described in chapter 10, 'Games'), must have their own terminology.[20]

Ordinal numbers occur, therefore, much more frequently than appears at first sight. It is not only that such words as *mokuyōbi* (Thursday) are essentially ordinals, but that the same is true, where the context requires, of such words as *hasami* (scissors) when it refers to a position in *janken.* Wherever a number denotes one specific member of a recognised closed set, it will be an ordinal, so that, in the modern world, telephone numbers are ordinals, as are also the names of subscribers as listed in the telephone directory.[21]

If ordinal numbers must always have a specific *denotation*, they may also have any number of *connotations*. The meaning of 'Thursday' is extended by 'Thursday's child', just as, 'Thursday' extends the meaning of 'four'.[22] Connotations of different numbers recognised by the Japanese will constantly recur throughout this book.[23]

The process of ordering, inherent in the use of ordinal numbers, is essential for counting. Given a series of number words, such as the Japanese *ichi, ni, san* ..., the recital of these words coupled to the selection, one by one, of the members of a set, is what the process of counting consists of. The denotation resulting from this process is adventitious, for although one member of the set must be selected as the first to be counted, and as such to correspond to *ichi*, and then another as the second, to correspond to *ni*, and so on, the final number, whatever it is, would have been reached by selection in any other order. This, the case of *vicariant* order, is the basis of the transformation from ordinal to cardinal numbers. It establishes the number of a set as something independent of the order of the members and, *a fortiori*, of their separate identities.

At this point, the essential character of the cardinal number, as a complete abstraction, becomes clear. The numeral, as a symbol representing a cardinal number, faces in two directions. Ontogenetically, it emerged from the linguistic process of counting applied to the members of a set, selected in a certain order. Logically, the numeral represents a perfect abstraction, called 'number', which is but one member of an infinite class, capable of combining, operationally, with all other members, according to purely formal rules: at least at an elementary level the system constituted by these rules is called arithmetic,[24] or in Japanese, *sūgaku*, or 'number science'. Because the logical theory of arithmetic does not allow the set of cardinal numbers to be closed, number words, or numerals, will always be an imperfect, because essentially incomplete, representation of cardinal numbers (Crump 1990: 10). The Japanese, in particular, have gone a long way in avoiding the embarrassment of working with an infinite series of numbers by developing alternative number systems, based upon finite sets, accepting in such cases the need to modify the formal rules of arithmetic.[25]

The scope of arithmetic is greatly enhanced by good written notation combined with an efficient means of calculation. In Japan the traditional written notation was confined to the resources of the Chinese characters, or *kanji*, which I describe in chapter 2. The defects inherent in the use of *kanji* were, however, largely overcome by the use of the abacus, as I describe it in chapter 9. None the less the Japanese remained preoccupied with elementary arithmetical constructs, which from the sixteenth century onwards were developed under the name of *wasan*, although this was never more than a mathematical recreation of certain classes of samurai.

Finally, if the scheme set out in this section assumes a sort of cognitive ontogenesis, based on the order, ordinal numbers, natural numbers, cardinal numbers, I do not intend this to mean that this view reflects how the Japanese see the question. In practice, given their obsession with numbers, most Japanese would probably incline to the view that 'the series of numbers is an innate intuition, present at birth in all members of the human race'. If, then, at first sight, it would seem that in relation to numbers no answer need be given to Lancy's question, 'at what point does genetic evolution stop and cultural evolution take over in managing the development of cognition' (Lancy 1983: 142), Japanese culture would still claim that its own use and understanding of numbers was unique. I shall, in this book, give any number of instances which show that this claim goes far beyond the historical achievements of *wasan*. None the less the fundamental principles developed in this section still apply, where context so requires, even in Japan.

## THE TAXONOMY OF NUMBER USE

The range of possible uses of number depends not only upon the distinctive properties of ordinal, natural and cardinal numbers, as they are recognised within any given culture, but also upon the scope which language allows for their application. In the context of number, ordinary language, whether spoken or written, is not sufficient to represent all the different ways in which numbers, and numerical operations, can be expressed. Numerical phenomena, such as music (even in the elementary form of the chiming of a clock), can communicate far outside the range of ordinary spoken language. The limitations of written language are even more formidable: long before the computer age, Japanese felt at home with such instruments as the abacus, which represented numbers in visible form so as to allow for operations for which the written forms of numbers were quite unsuited. The difficulties encountered in writing about a complex, heterogeneous, wide-ranging numerical culture, such as that of Japan, are largely due to the limitations of language in dealing with numbers. In the particular case of Japan these are compounded by the very considerable differences between Japanese and English, whether spoken or written. For this reason chapter 2, which follows this introduction, is devoted to the way in which numbers are incorporated into the Japanese language. Chapter 3 is concerned not with ordinary Japanese, but with what may be called 'meta-languages' which are constantly used for what are essentially numerical purposes. The alternative number systems, which have their own distinctive vocabulary, cannot even be used for straightforward counting, let alone for the processes of ordinary arithmetic, but that was never intended. In only one case (that of the so-called *kanshi*) is there a base of 10, and then it occurs only in combination with a base of 12. The practical application of numerical *meta-languages* is to be found above all in the field of fortune-telling, which provides the subject matter of chapter 6.

Where chapter 2 is largely concerned with essential questions of syntax and morphology, chapters 4 and 5 look at the ways in which numbers occur, idiosyncratically, in everyday language use. Chapter 4, which is concerned with proverbs and aphorisms, can be seen as a specifically Japanese instance of a phenomenon familiar in many other cultures: this is the use of numbers to express, in shorthand form, some general precept, applicable in a wide range of different contexts. The meaning of the Japanese *nichō isseki*, both literal and figurative, is immediately clear from the English translation, 'two birds, one stone'. Chapter 5, on the other hand, is concerned with two types of number use which are quite distinctively Japanese. The first, the incorporation of numbers in proper names, if known in the west, occurs there on nothing like the same scale. The second, the use of numerical formulas to determine which names are auspicious in a given context, depends entirely on the idiosyncratic form of written Japanese: this, the science of *seimeigaku*, has no western equivalent.

Chapter 8 extends the principle of communication, by means of numbers, beyond the realm of language into that of the use and organisation of space. This is not only a question of conscious design, whether in art, architecture or the lay-out of gardens, but also one of interpreting landscape and topography according to numerical principles.

At this stage, the way numbers are used is determined, in increasing measure, by their inherent arithmetical properties. An example, which I consider in greater detail in chapter 8, is provided by the well-known Kyoto temple of Sanjūsangendō. The building of this temple by the Emperor Goshirakawa over eight hundred years ago was an event which established the number 33, in its Japanese

version, *san-jū-san* in the local culture.[26] Although it would be absurd
to suggest that this was anyone's first experience of this number, it is
still important to note that this particular occurrence, however
novel,[27] recalled to the citizens of Kyoto not only other occurrences
of the same number but also its own distinctive arithmetical
properties. That is, it was accepted that the number 33, just as any
other number, had some sort of existence in its own right. This means
that the number, in popular cognition, was not established as a sort of
common denominator of all the instances in which it occurred (of
which *Sanjūsangendō* for a brief period of time was but the most
recent), but as something, existing in its own right and imposing its
own peculiar characteristics on anything to which it happened to
apply.

This being so, two questions arise. The first is: what are the
peculiar characteristics of the number 33? The second question is:
what is it that this number actually applies to? It is helpful to consider
the first question quite intuitively, first from the point of view of any
modern western culture, and then from that of the particular version
of oriental culture which is to be found in Japan.[28] In the west the
number 33 may be seen as being odd, composite (i.e. being 3 x 11,
not prime) and containing two identical digits (indicating that it is
divisible by 11). A mathematician could notice that 33 belongs to the
class of numbers which can be expressed as the sum of three squares
(so that $33 = 4^2 + 4^2 + 1^2$), or even that it exceeds by 1 a power of 2
(so that $33 = 2^5 + 1$). This means that the number 33 expressed in its
binary form, 100001, clearly belongs to the class of what Gerschel
has defined as that of the *nombre marginal* (Gerschel 1962: 696).

This is the point at which the purely arithmetical properties of a
number become culturally significant. Gerschel's marginal numbers
do not necessarily have the property that they exceed by 1 a power of
2 (such as 32) or a power of 10 (such as 1,000), but in every
case in which such a number occurs, it does exceed by 1 the number
of members contained in a closed class recognised in the local
culture,[29] and this is what determines its cultural significance.[30] In the
case of Japan, the definition of marginal numbers on the basis of their
representation in binary form is a well-established tradition. The
importance of such representation is to be found in two essentially
numerical institutions, neither of Japanese origin, both of which are
also familiar in the west. These, the *mandala* and the *yin–yang*
opposition, are among the alternative number systems which I
describe in chapter 3.

At the point now reached, any number, expressed in binary form, is seen as connoting something completely extrinsic to its intrinsic, arithmetical properties. Returning to the case of Sanjūsangendō, the arithmetical property of the number 33 – that it exceeds by one a power of two – is extrinsic to the meaning that led this number to be chosen, both for naming the temple, and determining the form of its ground plan. From the purely linguistic point of view, the word *sanjūsangendō* could be taken to be an arbitrary *sign*, representing this particular temple.[31] This is how most residents of Kyoto regard it. None the less, in contrast to most of the other names given to the many temples to be found in the city, *sanjūsangendō* was always meant to convey a particular message, relating to the number 33. The arithmetical significance of this number has already been considered; it is to be noted that this governs not so much its choice in relation to both the name and design of the temple, as its occurrence in the Lotus Sutra, which provided the original inspiration for building it.

The instance of the number 33, analysed in the previous paragraph, is an example of metonomy, for the inherent properties of the meaning of the sign – in this case the number 33 – are taken to designate an attribute in a quite different cognitive realm. The arithmetical perfection inherent in the number 33, represented by its binary form, 100001, is of different order from the *mystical* perfection of the Kannon Buddha, to whom the temple of Sanjūsangendō is dedicated. The equation of the two is a cultural construct of a kind which constantly recurs in Japanese culture.

A cardinal number is not only significant for its inherent arithmetical properties. Operationally, the way in which cardinal numbers can combine with each other is much more important. This is not only a question of the applying familiar arithmetical operations, such as addition and multiplication, but of extending the process of abstraction into the realms of algebra, where numbers are conceived of in entirely general terms. This process enables formulas to be expressed for solving general problems, at the same time allowing for the possibility of extending the class of numbers beyond that of positive integers or whole numbers isomorphic to the series of natural numbers. The first step in this direction, which is to extend the class to include negative integers, may be simple enough, but the next step, requiring the introduction of fractions, is not as elementary as it would seem at first sight,[32] and although as chapter 2 shows, the Japanese have no difficulty in expressing fractions, it is still significant that units of measurement are defined so as to greatly reduce the need to use fractions in daily life.

Irrational, let alone imaginary or complex numbers, were hardly known in the pre-modern period, although they could well occur, conceptually, as solutions to quadratic equations. In practice, the Japanese were little interested in the scope of algebraic equations for widening the definition of the concept of number. The autochthonous Japanese algebra, which was an integral part of *wasan*, took the form of mathematical recreations or puzzle-solving, so that although the problems stated were highly involved, they always lacked depth (Sugimoto and Swain 1978: 275).

The original Japanese contribution in the field of applying mathematics to the natural sciences was equally insignificant. Although astronomy in particular was as important as it was in China, the methods used were all of Chinese origin. If in the pre-modern period the Japanese broadened their scientific base, it was by adopting the methods of *rangaku*,[33] or European science as it was known to the small Dutch community on the island of Deshima in the bay of Nagasaki.[34] The reforms of the Meiji era denied all recognition to *wasan*, which was relegated to the status of a hobby for antiquarians.[35]

If at the present day the tradition established by *rangaku* dominates the teaching and application of mathematics, even to the point that no successful Japanese alternative to western computer languages has emerged, it is none the less remarkable that in the field of consumer electronics Japanese products dominate the world market. I consider, in chapter 11, whether this is simply the result of superior production and marketing strategies (which are certainly not confined to this field) or whether, in some way, the Japanese are peculiarly gifted in their use of numbers.

The answer to this question may be found in the realm of formal, numerical systems, such as provide the basis for traditional institutions, which, at the level of popular culture, are not conceived of as essentially numerical. Music, and even more so games, are characteristic institutions of this kind. The Japanese appear to be fascinated by almost any game with a mathematical base, whether, like *go* and *shōgi* it belongs to an ancient domestic tradition or, like golf and baseball, it is a recent import. Japanese children playing *janken* are taking the first step on the way to becoming indoctrinated into the games-playing culture which I describe in chapter 10.

Such then is the scope of my book. I have tried, where possible, not only to describe and interpret any number of separate instances of the Japanese use of numbers, but also to explain their significance for those who wish to understand the way Japanese think and act. In the result it will be seen that numerical institutions impose on many different aspects of Japanese life a formal structure which determines the way in which problems are conceived, then going on to provide the means of solution.

This point can be illustrated by the way in which 'risk' is conceptualised. No single Japanese translation of the English word is adequate. A risk which cannot be measured is seen as *kiken*, which is otherwise seen as *danger*.[36] But once a risk can be expressed in terms of *kakuritsu*, or 'measure of certainty', it is somehow defused and assumes more the character of *bōken*,[37] a venture which may be expected to lead to gain. Seen cognitively, the basic idea of *kakuritsu* is numerical, and numbers, although they have an integrity of their own, can be mastered. A formal, numerical basis is then established for managing risk, and this makes possible the institution of insurance, a type of business in which the Japanese are extremely proficient. Insurance, as a formal economic institution, based on actuarial mathematics, is very recent in the history of Japan. The underlying principle, that numbers, properly used, provide the basis for establishing order, is much older, and in this book I shall give countless instances of its application.

# 2 Numbers in the written and spoken language

## LANGUAGE AND NUMBER

The series of numbers 1, 2, 3 ... in English correspond to the words *one, two, three* ... used in counting. It is these *number words* which embed numbers in language, so as to provide the only historical context in which they could have originated as part of human cognition. Language means in the first place spoken language, which – from the neurophysiological evidence provided by palaeontology – has probably been part of human communication for some half a million years. Writing (which for the first time provided good historical evidence for the development of language) first appeared somewhere in the middle east about 10,000 years ago. It is highly significant that the first writing was developed so as to record quantities measured by numbers. This shows that at least some of the languages spoken before this time contained number words, although even today there are still languages in Australia, and perhaps elsewhere, which contain no numbers.[1] In the case of Japanese, in which both the use of written numbers, or *numerals*, and the historical record begins with the first contact with Chinese culture some 1,600 years ago, there is clear evidence of the existence of number words in the language spoken before this time: such words are still part of Japanese, although their form has certainly been modified over the course of time.

If the series of natural numbers is unlimited (a proposition accepted in the western world since before the time of Plato), the same cannot be true of the number words in any language. There must be a limit to the process of counting, simply because no vocabulary can contain an unlimited number of words. Many languages containing such words go no further than some number less than 10, but in all such cases the series of number words is continuous in the sense that there will be words for all numbers up to the limit imposed by the lexicon.[2] If, for instance, this limit is 5, then there will always be a word for 4, and so on down to 1 (although the status of 1 as a number is often equivocal).[3]

In the formation of number words two general principles apply to almost all known languages. The first is that low numbers, up to a recognised limit, are represented by words, which, systematically, are quite unrelated. In English this is certainly true of the word for numbers up to 5, if not up to 10.[4] The second principle is that almost all numbers, above another limit, if they exist at all, will be formed by some principle of composition. In English this limit is 10, so that 'sixteen', for instance, contains two components, meaning 6 and 10. In any language, this second limit cannot, *a fortiori*, be lower than the first, but there are certainly languages in which the two limits do not coincide, so that the numbers falling between them are sometimes composite and sometimes not.

The lexical principles applying in the realm of *composite* number may be quite chaotic,[5] particularly for relatively low numbers. The numbers up to 100 in Indo-European languages provide such examples as *quatre-vingt-douze* for 92 in French, but Yoruba, one of the most important languages in West Africa, is far less ordered. No matter: at a certain point order must be imposed if larger numbers are to be represented. In French, once one has reached 100, there are no more than quite minor problems. For the middle range, generally between 10 and 100, where numbers are composite but subject to disorder, the countless different known languages can be ordered on a scale, with the simplest cases at one end and the most complex at the other. Chinese is a simple number language *par excellence*, where Yoruba is at the opposite extreme. The point is important since command over the middle range is important if numbers are to be used *operationally*, that is, in ways which make use of their essentially arithmetical properties.[6] It is at this level also that visual representation – not necessarily by writing – of numbers enormously increases their operational utility.[7]

The hallmark of a simple system for composing number words is simply to be found in an economy in the rules governing the process. The first step is to establish a base number, such as 10,[8] which in successive powers, 100, 1,000, and so on, provides a *frame* in which to construct number words. This process plainly cannot go on for ever, and at some stage the normal counting process, although theoretically possible, no longer provides the ordinary means for forming numbers.[9] In English the word 'million' is never used in communicating a telephone number: even the year, 1992, is seldom spoken out in full. At a certain level also, numbers in the frame will almost certainly be composite: in English this process begins with 'ten thousand', whereas in Japanese it does not begin until *jū man* (hundred thousand).[10] The series of low numbers below the base number, which on a system with base 10 will be the numbers 1 to 9, then provides a *cyclic* system, which, in combination with the frame numbers, allows all numbers likely to be used in spoken discourse to be expressed by a word without any ambiguity, although in some languages such as Japanese there are alternative words for some numbers.[11] In a simple system, such as the Chinese, little more need be said, save that the number 1, as a member of the cyclic system, is generally suppressed when it governs a frame number, just as in English 10 is 'ten' and not some word derived from 'one' and 'ten'. This process, known as *one-deletion* occurs in a great many languages. The general absence of a zero in such systems – at least when they originated – means that frame numbers, governed by a zero, are simply not part of the number word. For example, 207 is spoken as 'two hundred and seven', without any explicit indication of the missing tens.

Using the linguistic term *morpheme* to mean any non-reducible element in a language with its own recognised meaning, then the number words will be composed of a very limited number of morphemes:[12] in English some twenty are sufficient to represent every number up to a million. These will comprise all the cyclic and non-composite frame numbers, which, being capable of standing alone, are *unbound* morphemes, together with such *bound* morphemes as '-ty' and 'thir-', which only occur in a composite number word. In English number words, 'and', as in 'two hundred *and* seven', must be reckoned as a bound morpheme, whereas in this particular case the other components are unbound.[13]

Now, if the English case is simple it could be even simpler, as is shown by the case of Chinese, in which nine cyclic number words and four frame number words, all unbound morphemes, are sufficient for every number up to a hundred million. It contains no bound morphemes, with a limited number of unbound morphemes combining according to a single rule so as to represent all numbers capable of being expressed. This is the hallmark of a simple system.

When it comes to the written forms for number words, the position is altered quite radically. In the first place, counting by means of written signs is possible simply by repeating the same mark, so that 1, 2, 3 ... can be represented by /, //, /// ... a process without any parallel in spoken language.[14] Such representation may be made more manageable by using a base, say 5, to establish recognised larger units.[15] This can be achieved by using a fifth stroke to round off the first four, so that one writes not 𝄇 but 卌. The Japanese use the character 正, whose five strokes are written in the following order: 一, 丅, 干, 乍, 正, so that 9, for instance, would be represented by 正 乍.

Right at the very beginning, therefore, the simplest written form loses contact with its spoken equivalent. It is moreover extremely difficult to re-establish contact, quite simply because conventions of the spoken language, such as one-deletion and zero-suppression, make any literal written representation extremely confusing. The reason for this is to be found in the characteristic uses of written numbers. So long as such use is confined to purely documentary transactions, such as payment by means of a bank-note or the keeping of archives such as census returns, the actual written form is not crucial. As soon, however, as numbers must be used operationally, as in the elementary arithmetic used in accounting, the written number system must be amenable to simple arithmetical rules, or algorithms, such as children now learn in primary school for addition, subtraction, multiplication and division.

Given the demands made of numbers in their written form, it is not surprising that we only know of one original case of a language which represents every numerical morpheme occurring by its own distinctive character, so that there is a perfect one-to-one correspondence between the written and the spoken versions: this language is Chinese.[16] Even then the original written numbers did not pass this test;[17] historically it is of decisive importance that at a certain stage the written forms were adapted so that they did so, and it is these forms which are still in use today, not only in China but in Japan. The transformation, which took place well over two thousand years ago, was only possible because of the utterly simple syntax of number words in the spoken language. Whatever the advantage of having such transparent written forms, the price still to be paid was that these were still not adequate for any but the simplest arithmetical operations.

To this problem there is only one solution, which is the visual representation of numbers by a place-value system. The best-known example of this is provided by Arabic[18] numerals, from 0 to 9, whose value in *any* numeral is determined by their place in it. This means that 2 in 207 is 200, simply because it occupies the third place from the right end, and *hundred* is the third component in the frame sequence *one, ten, hundred, thousand* and so on.[19] Once this convention is adopted, the divorce between written and spoken form is near complete, and simply does not matter. The fact that the French *quatre-vingt-douze* for 92 is perhaps frustrating if one is learning to speak French, but a French child has no more difficulty in dealing with 92 in a sum than an English child, for whom the spoken form 'ninety-two' is much more transparent. But what about a Chinese child, for whom the written form 九十二 will correspond precisely with the spoken form *jiǔ shì èr*. Can the written form be used in arithmetic. The practical answer is no, so that at the present day Chinese children are also taught to work with Arabic numerals, which of course fit the spoken language no better than they do French or English. At the same time the Chinese, for arithmetical operations, continue to use the abacus, a calculating instrument which represents the numbers occurring according to a place-value system isomorphic to that of the Arabic numerals. The actual form of numerals on an abacus is more elementary, since every number, from 0 to 9, is a combination of beads, with those below the bar representing units, and those above multiples of five. This is illustrated in figure 9.1 (see p. 128).[20] The lesson is that to deal with numbers at all efficiently, in arithmetical operations, there is no alternative to a place-value system. This is even true with electronic computers, with the difference only that they work on a base 2 and not 10.[21]

However the problem of representing numbers, whether in spoken or written form, is solved, the question still remains as to how linguistic factors affect their actual use. This of course is but one aspect of the subject of this book, the use and understanding of numbers, but it is a decisively important one, particularly in the extremely idiosyncratic context of modern Japan. Numbers, in any culture in which their operational use is at all advanced, must face in two directions, that of language on one side and that of arithmetic on the other. The two directions relate, if somewhat imperfectly, to the difference between the spoken and written forms and, in a rather different cognitive domain, to that between ordinal and cardinal numbers. In the English-speaking world convention dictates that a number is written out in full when its use is descriptive rather than operational: contrast the 'nineteenth century' with a '19p stamp'. In the case of Chinese, and derivatively that of Japanese also, this distinction is only possible if Arabic numerals are used, but such use was unknown before the late nineteenth century. In the English case it corresponds roughly to the distinction between language and arithmetic, although the case of the 19p stamp, as many others, faces in both directions. Those in the sorting office handling the letter can see from the face of the stamp whether the right postage has been paid, so that in this case the numerical information is descriptive. On the other hand, those working behind the counter, who sell the stamp, perhaps with others of the same or a different amount, and then have to work out the change for a bank-note offered in payment, are involved in an operation depending upon the *arithmetical* properties of 19. It is not for nothing that counter-clerks in Japanese post-offices have their own abacus to help them with such calculations. None of this is possible with the 19 in '19th century' except in a very contrived sort of way: 4 x 19, for instance, is not conceivable as an operation applicable in such a case.

In language the use of number words is primarily descriptive, so that, in English grammar at least, they may be seen as adjectives that qualify a noun.[22] The young child, as it learns to use numbers, must somehow marry the numbers it has learnt by counting with different categories of countable objects. Logically 'seven apples' is prior to an understanding of 'seven', however counter-intuitive this statement may be. What 'seven apples' represents is the intersection of the class of aggregates containing apples and that of collections with seven members. If we look at this latter class from the perspective of our own culture, we see that it is extremely heterogeneous. It includes not only the seven apples just counted, but the 'seven stars in the sky', the 'seven deadly sins', the 'seven days of the week' and the 'seven hills of Rome', together with countless other collections which one could conceive of or constitute *ad hoc.*

The cardinal number 7 is the only thing these collections have in common: linguistically 'seven' is the only modifier which applies to all of them, but it is still not an inherent property of any of them. Take away one apple from the pile, and neither that apple nor those remaining are changed in any way, but the property of being 7 is completely eliminated, and seven, as a modifier, no longer describes (except in the past tense) anything connected with the apples. If the property of being 7 in this one case is eliminated, the concept of 7 remains, like the smile on the Cheshire Cat, which, as Alice found, was the last part of it to disappear from sight (giving rise to all sorts of epistemological problems).

It is precisely this concept of 7 which is operational. It is no accident that the numerical quantities occurring in everyday arithmetic relate for the most part to abstract quantities, such as units of measure or money (which may be taken to measure *value*). Whatever concrete results may be reached in contexts in which arithmetical operations are used, the latter are abstract, autonomous and context free. Arithmetically there is no difference between calculating the price of 7 19p postage stamps and calculating the weight of 7 19lb sacks of potatoes.

This fact is particularly important in the context of Japanese, which does not allow numbers the *adjectival* property of applying to all nouns. In Japanese, the words for seven (*nana*) and apple (*ringo*) cannot directly combine to express 'seven apples', and Japanese is certainly not the only language subject to this restriction.[23] The problem is solved by the use of counters, or in Japanese, *josūshi*,[24] which *can* be governed by numbers, and so provide a sort of interface between any given number and any noun which that number must govern. This means the constant use of phrases such as the English 'seven head of cattle'. Thus 'seven apples' can only be expressed in Japanese with the help of the counter *ko*, so that the complete form is *nana ko no ringo*. In some cases the counter can also stand alone, so that *sannin* is simply 'three people', whereas *sannin no kodomo* is 'three children', or *sannin no sensei*, 'three teachers' – *nin* being simply the general counter for human beings,[25] leaving aside such special cases as Shinto gods, who have their own counter. In the case of the measurement of physical dimensions, time and money, the appropriate units generally need no counter, so that fifty (*gojū*) Yen (-*en*) is simply *gojūen*. Having said all this, there are still special rules for using the autochthonous Japanese numerals (which have not been introduced) in forms which count up to four people and up to ten days, or sometimes more.

Japanese, therefore, makes explicit the logical priority of 'seven apples' (or seven of anything else) over the number 7. This part of Japanese cognition is reinforced by the use of the abacus, since the numbers occurring in any calculation count the beads used in it.[26] The abacus therefore represents the sum $7 + 11$ as 7 beads added to 11 beads, and indeed abacus calculation is simply known as *shu-zan*, meaning 'bead calculation'. The secret of success is to be found in the use of a place-value system for ordering the beads so as to represent any number occurring, for the written form of numerals, being ordered according to a different system, cannot be used directly for calculation. The price paid for this success is that a number, say 7, can only be abstracted from 'seven apples' by being moved to seven beads. This is a matter of cognition, not of arithmetical skill, which in practice does not suffer from the limitations of the Japanese language. Indeed in everyday counting and arithmetic numerals are used *perfectly correctly* without counters.

Finally, there is the matter of the distinction between cardinal and ordinal numbers. In terms of word formation, the Japanese process for deriving ordinal from cardinal numbers is very simple, as it is in many other languages.[27] The simplest rule is to add the suffix *-bamme* to any number word, so that *san-bamme* (third) is formed from *san* (three). In governing a noun, once again the particle *no* is used, so that 'the third apple' is *sanbamme no ringo* (noting in this case that the counter *ko* is not needed). In the case of numbers up to ten, the autochthonous numeral system can be used, simply with *-me*, so that *mittsume* for 'third' is also possible. An alternative is to use the prefix *dai-*, so that *daisan* also means 'third'. This form, in conjunction with *ichi* (one), often occurs in contexts where *dai-ichi* refers to the 'number one selling brand',[28] which fits in also with the common *ichiban*, the normal word for 'most'.

The ordinal number, much more than the cardinal, is an adjective describing an inherent and not an adventitious property of the noun it governs. February is by definition the second month of the year. Japanese, much more than English, is inclined to make such meaning explicit, so that February becomes *ni-gatsu*,[29] literally 'two month'.[30] Such use is particularly characteristic of closed sets, such as that of the twelve months of the year, which as a series of ordinal numbers has a beginning and an end, which as a series of ordinal numbers forms as 'first' (which lexically is not derived from 'one'), and 'last', corresponding to the Japanese *hajime no* and *saigo no*.[31] (Alpha and omega are the primordial ordinals which enclose the whole set.) And although it is true that cyclic sets repeat themselves, and in doing so allow the process of counting by number words to continue – so that every time January comes round, the year moves on by 1, with 1991 succeeding 1990[32] – even then both dates, as indeed any date, are essentially ordinal numbers; 1990 is the 1990th year from a recognised starting point. The Japanese do not have to count so far, since they start again with every new emperor – so that the present year is Heisei 3.[33]

At this somewhat esoteric level there is a sort of litmus test for distinguishing cardinals from ordinals. A cardinal set is isomorphic with any other cardinal set containing the same number of members, regardless of the order of members of either set. The order of such a set is said to be *vicariant* (Piaget 1952: 96). Inherent in this definition is the property that a new member may be inserted at any place in the set with the simple result of increasing its number by one. This property, which inheres in any cardinal set, however large, corresponds to the proposition that there is no limit to the series of natural numbers. (There is never a last straw which breaks the camel's back.) An ordinal set is not vicariant, and it is this property which allows it to be closed. The present time confronts us, in increasing measure, with ordinals which disguise themselves as cardinals, because the property of being closed is no longer transparent. A telephone number, for instance, is an ordinal, but whereas any Japanese knows that there are twelve months in a year, not even a telephone company executive could say precisely how many telephones there were in Japan. The position here is made even more opaque by the fact that lists of telephone numbers (in directories) are based upon an ordinal principle which is lexical rather than numerical (although the telephone company no doubt maintains lists in numerical order for use by its own accountants and engineers).[34] Applying the litmus test once more, one must realise that however much Japan may become saturated with telephones – a process which seems to be well under way – the actual number will always remain finite. And last but not least, the telephone number, save possibly out of pure perversity, is never subjected to an arithmetical operation; just imagine how subscribers would react if their monthly rental were based upon the square root of their telephone number rounded off to the nearest integer. The true merit of ordinal number words, as a linguistic device, is their unique and systematic precision: there could be any number of Japanese with the name Nakamura Takeshi, but only one with the telephone number 075-5941994.

The ordinal property of closed sets means that an ordinal number word based on a number larger than that of the number of members in the set can have no meaning according to the normal definition. In Japanese, *jūgogatsu* is a sort of non-word, because 'fifteen-month', the only meaning it could have, is not an actual month. One might expect a meaning to be given to *jūsangatsu* – or 'thirteen-month' – as representing the intercalary month added every three years to correct for the extra days in any year after the lapse of twelve lunar months, but in fact one of the existing months is repeated, although the festivals occurring in it do not take place a second time. Where such a *marginal* number is recognised, it has a special meaning, such as that of the nineteenth hole in golf, which refers to the clubhouse bar. Japanese cases will also occur, but *jūsangatsu* is not one of them.

## JAPANESE ORIGINS

The status of the Japanese language is beset by paradox, particularly for the Japanese themselves. The ideal, maintained by cultural patriots, is of a basic language, spoken by their earliest ancestors in the Japanese islands, and uncontaminated by any outside influence. This ideal is part and parcel of the concept of 'Yamato', a word which connotes everything which is essentially Japanese.[35] The *kanji* which represents this word has another form, *wa*, which is used as a suffix with much the same positive meaning. Thus *wabun* is the basic Japanese language, just as *wasan* (introduced in chapter 1) is the basic Japanese arithmetic.

The concept of a pure and uncontaminated Japanese is difficult to maintain in the face of linguistic, archaeological and historical evidence,[36] which even the cultural patriots cannot completely disregard. The linguistic evidence is that Japanese is not an isolated language, such as Basque, but a member of the Altaic family,[37] which is represented by such other languages as Korean at one geographical extreme and Turkish at the other. The archaeological record supports this conclusion: the Japanese were an immigrant population, with a substantial component from the Asian mainland, who largely displaced the original Ainu inhabitants. But all this the hard-core Yamato enthusiasts can brush aside. Everything happened so long ago, at a time before there were any historical records, that an alternative mythology provides an equally valid charter for Japanese traditional culture. The gap is considerable: any other Altaic language is about as close to Japanese as, in the Indo-European family, Armenian is to English. The relationship, if it can be proved to the satisfaction of scholars, means nothing to the man in the street.

The historical evidence is of a quite different order of significance. There is nothing in the record which is not based on the continued use and adaptation of different versions of Chinese, both written and spoken, over a period not all that far short of two thousand years. Whatever there is in Yamato can only be seen in a mirror made in China.[38] But for all that, there is something to be seen, even in the numbers used in daily life in twentieth-century Japan. These autochthonous numbers, in their modern form, are:

| | | | |
|---|---|---|---|
| *hitotsu* | 1 | *muttsu* | 6 |
| *futatsu* | 2 | *nanatsu* | 7 |
| *mittsu* | 3 | *yattsu* | 8 |
| *yottsu* | 4 | *kokonotsu* | 9 |
| *itsutsu* | 5 | *tō* | 10 |

The remaining numbers in old Japanese are no longer used, except occasionally in proper names (which can then be difficult to decipher)[39] and in one or two common forms, such as *yaoya* – literally 'eight-hundred-shop' – for a greengrocer. Before looking at the actual use of the numbers up to ten, which one hears in everyday conversation (and particularly in counting up to ten), it is worth noting how, by a vowel shift, *hitotsu* (1) becomes *futatsu* (2), *mittsu* (3) becomes *muttsu* (6), *yottsu* (4) becomes *yattsu* (8), and rather less obviously, *itsutsu* (5) becomes *tō* (10), the derivative form being in each case twice the original form. This type of number word formation, if rare in language generally, causes no problems to the Japanese. The use of these autochthonous numbers, although very common, is still restricted by the use of syntax. In particular they cannot combine with *josūshi*, so that for instance *nanatsu* cannot as such be used in a phrase meaning simply 'seven apples'. On the other hand, a sentence like, 'I bought seven apples at the greengrocer's' can well be translated in the form, *ringo o nanatsu yaoya de katta* – a circumlocution meaning 'apple (*accusative*) seven greengrocer at (I) bought'.

In two special cases, both very common, there are composite forms based on the authochthonous number words. The first is quite simple. The counter *nin* for people, with 'one' and 'two', is supplanted by *hitori* (in place of *ichinin*) and *futari* (in place of *ninin*), so that 'two teachers' is for instance *futari no sensei. Yottari*, instead of the more common *yonin*, is also used for 'four people'.[40]

The second special case is much more involved. In counting days, particularly for use in dates, the following autochthonous forms are used:

| | | | |
|---|---|---|---|
| 1st | *tsuitachi* | 6th | *muika* |
| 2nd | *futsuka* | 7th | *nanoka* |
| 3rd | *mikka* | 8th | *yōka* |
| 4th | *yokka* | 9th | *kokonoka* |
| 5th | *itsuka* | 10th | *tōka* |
| | | | |
| 14th | *jūyokka* | 20th | *hatsuka*[41] |
| 24th | *nijūyokka* | | |

There is also a special form, *hatachi*, for '20 years (of age)'. The remaining numbered days are formed with the suffix *-nichi*, used as a normal counter. The difference in use can be seen in the translation of 'May the third' and 'May the thirteenth'; the former is *gogatsu mikka*, the latter, *gogatsu jūsannichi*.

## THE CHINESE CONNECTION

The autochthonous numbers, introduced in the previous section, are incomplete, at least in their present-day usage.[42] This is simply the result of the original system being supplanted by number words brought from China, which form as complete a system as ordinary use could ever require.[43] Since, also, there is a perfect one-to-one correspondence between the spoken and the written forms, these are now given together in table 2.1, which shows the versions now incorporated into Japanese.

The rules for using this system are simple. In representing any number, the powers of 10 are taken from the frame system, descending from the higher order – just as in English. Those with coefficient zero are left out, and one as a coefficient is deleted except, commonly, in the cases of *ichiman* (for 10,000) and *ichioku* (for 100,000,000).[44] *Issen*, for 1,000, also occurs occasionally. This is simply the process of *one-deletion* mentioned on page 16. With the exception of the numbers from 11 to 19 (which in many languages are a *marked* or special case), the system is essentially the same as that of spoken English. What makes it remarkable in the case of Chinese, and derivatively, that of Japanese also, is that the written form corresponds so precisely to the spoken: this is not a property of any other known system of number words (Menninger 1977: 458). In principle, the Japanese could completely give up the use of their autochthonous system, for that imported from China is adequate for all purposes, as it is, naturally enough, in the different versions of Chinese.

The Chinese written numerals are but one example of the Chinese characters, known as *kanji*, used in written Japanese. The use of the *kanji* numerals illustrates two distinctive features of Japanese, which have no parallel in any version of Chinese. The first of these is the possibility of alternative readings for a given character; the second, the phonetic representation of words, or parts of words, by means of one of the two *kana* syllabaries.

In Chinese a given character can only be pronounced as a single monosyllable, which is either a word in itself or a recognised part of a longer word.[45] The former case includes all characters representing numbers, so that in Mandarin 二 is pronounced *èr* and means 'two', just as 百 is pronounced *băi* and means 'hundred', the two combining to form 二百, pronounced *èr-băi* and meaning *two hundred*, or in reverse order, 百二, *băi-èr*, meaning *hundred and two*. Now all this can be done with the Japanese *ni* and *hyaku*, meaning respectively

*Table 2.1* Chinese numerals

(a) *The cyclic system*

| One | Two | Three | Four | Five | Six | Seven | Eight | Nine |
|---|---|---|---|---|---|---|---|---|
| 一 | 二 | 三 | 四 | 五 | 六 | 七 | 八 | 九 |
| *Ichi* | *Ni* | *San* | *Yon* *Shi* | *Go* | *Roku* | *Nana* *Shichi* | *Hachi* | *Kyū* *Ku* |

(b) *The frame system*

| Ten | Hundred | Thousand | Ten thousand | Hundred thousand | Million | Ten million | Hundred million |
|---|---|---|---|---|---|---|---|
| 十 | 百 | 千 | 万 | 十万 | 百万 | 千万 | 億 |
| *Jū* | *Hyaku* | *Sen* *Chi* | *Man* | *Jūman* | *Hyakuman* | *Senman* | *Oku* |

'two' and 'hundred'. Such a reading of 二 and 百 is known as an *on* or sound reading, the implication being that it is based upon the true *Chinese* pronunciation of the *kanji*. In Japanese almost all[46] *kanji* have such an *on* reading, simply because they came from China, and brought with them, to be incorporated into Japanese, the Chinese word which they expressed. The Chinese numerals are a case in point: all the *kanji* numerals have an *on* reading, which can be taken to be the normal pronunciation. However, in addition to the *on* reading, a *kanji* can have another, known as the *kun*, or explanation, reading, so that 二 ('2') is not only *ni* but *futatsu*. In a sense this reading explains the meaning of a Chinese character in Japanese. It is as if, sometime in the third or fourth century, a class of primordial Japanese were being given a lesson in *kanji*, and were told that 二 , while pronounced *ni* by the Chinese,[47] actually meant *futatsu*. In practice, the Japanese child learning the *kanji* numerals in the first year of primary school, already knows the correct use of *futatsu* and *ni*, without being conscious of either having any sort of ontogenetic priority. The fact that 二 is used in writing either word merely underscores the fact that they have the same meaning.[48]

The *on* and *kun* readings are not merely transformations of each other. In the general case, which includes also number words, context allows only one out of all possible readings.[49] There is however a common, though not a necessary distinction, between the *on* and *kun* readings, arising from the fact that the former represent words of Chinese origin which cannot take directly any of the wealth of agglutinative suffixes required by Japanese grammar, whereas the latter, being autochthonous are not subject to this restriction. Now since these suffixes correspond to nothing in Chinese, they are not at all readily represented by *kanji*, so that in normal texts they must be represented by characters from one of the *kana* syllabaries.[50] Each of these contains forty-seven characters in current use, every character (except that representing a terminalo *n*)[51], corresponding to a syllable in spoken Japanese, consisting of one of the five vowels, *a, i, u, e, o*, preceded at most by a single consonant. In *hiragana*, which is the syllabary in general use, the character つ is pronounced *tsu*, so that *futatsu*, for 2, can be written 二つ. Then, wherever this occurs, the reading is clearly *futatsu*, and not *ni*, so that the reader immediately knows which reading is correct. Such usage, although quite standard, is not obligatory, so that although one would normally write りんごを二つ八百屋で買った for the Japanese sentence given on page 24, still, if the *hiragana* つ were omitted, the *kanji* 二 would none the less be pronounced *futatsu*, simply because this is right in the context. In the case of the *kanji* versions for the months of the year, the distinction is critical because of the month-counter, *ka*, almost invariably represented by the *hiragana* か. Without the counter one has simply the name of the month, so that 五月, *gogatsu*, is 'May', where 五か月, *go*-ka*getsu*, means 'five months'.

## FRACTIONS AND MEASUREMENT

The decimal system of Chinese numerals also provided a structure for measurement. The metric system, established in the western world only in the last two centuries, was foreshadowed in China more than three thousand years ago. The basic unit of Chinese measurement is represented by the character 分, pronounced *fēn* in Mandarin. As a verb it means 'divide, distribute, distinguish', which all related to its use as a noun, meaning in the first place 'a fraction', or more specifically 'a tenth', and then according to context, a unit of length, of area, of weight, of money, and finally of time, where it represents a minute – in this unique case not a tenth, but a sixtieth part of an hour.[52]

The same character, in Japanese, has a *kun* reading, *wakeru*, which represents a transitive verb with much the same meanings as those given above for Chinese. There is also an intransitive form *wakaru*, which most commonly means 'to be known' or 'to be understood'.[53] The nouns represented by 分 have most commonly the *on* reading, *bun*, which, in the present context, can be taken to mean 'part, share' or 'degree'; *fun*, 'minute' and *bu*, 'rate' or 'percentage'. Both *fun* and *bu* occur in the traditional decimal system of weights and measures, which, although imported from China, was not nearly so comprehensive in its use of the concept represented by the Chinese *fēn*.[54]

The word for 'time' in Japanese is *toki*, a *kun* reading of the *kanji* 時. This same *kanji*, in its *on* reading, *ji*, combines with *kan* to form the word 時間, *jikan*, meaning 'hour', or more generally 'period', which is much the same reading as 間, *kan* (or in its *kun* reading *aida*), has when standing alone. *Ji* also combines with Chinese numerals to tell the time, so that, for example, 'six o'clock' is simply *rokuji*. The rule also applies to state the minutes, *fun*, so that 'five past six' is simply 六時五分, *rokuji gofun*, and 'quarter past six', 六時十五分, *rokuji jūgofun*. Only for 'half past' is there a special word, *han*, so that 'half past six' is 六時半, *rokuji han*. The minutes before the hour can be given simply by appending the word, *mae*, meaning 'before', so that 'five to six' becomes *rokuji gofun mae*, and 'quarter to six', *rokuji jūgofun mae.*

The use of 分, *bun*, to express fractions is equally straightforward once it is realised that the English order of numerator and denominator is reversed. *Five-sixths*, therefore, is 六分の五 , *rokubun no go.*[55] Once again, *han* (half) is the only exception to the rule.

The syntax of number words could hardly be simpler than it is in Japanese. At the same time, because what numbers represent is easier for the eye than the ear to grasp, the elementary representation of numbers in their written form is critical for understanding them. This is doubly true where the written language is as complex as it is for the Japanese.[56] The use of *kanji* embeds the numerals in the written language in a way which has no parallel in any western language. They are the most perfect exemplification of the principle that a *kanji* represents meaning.[57] In the use of *kanji*, numbers, while not a special case, still stand out for their simplicity; they are the most accessible part of the cognitive domain. It is not therefore surprising if, by one expedient or another, their range is extended throughout Japanese popular culture in ways largely foreign to any western culture of the present time.[58] If the Japanese are so much at home with numbers, it is largely because of their place in the Japanese language, both written and spoken.

## THE ORDERING OF *KANA* AND *KANJI*

The need to reduce both the *kana* syllabaries and the thousands of *kanji* to some sort of lexical order makes it necessary for the Japanese to work with numbers in ways which have no equivalent in the western world. *Kana* is in principle a simple case. There are five vowels in the order *a, i, u, e, o*, which may either stand alone (generally at the beginning of a word) or be preceded by one of nine consonants, in the order *k, s, t, n, h, m, y, r, w*, making fifty possible sounds to be represented by the appropriate characters. This is the basis of the *gojūon* 五十音, or 'fifty sounds', which for *kana* has the same meaning as the alphabet for western orthographies. As always in Japan, nothing is quite as simple as it looks, since some possible combinations are missing, and there is also a terminal *n* with its own *kana*. There is also an alternative order to that of the *gojūon* (which essentially establishes a two-dimensional matrix) and that is the *I-ro-ha*, an ancient poem in which all the sounds of the *gojūon* are represented, but none more than once. This is not only learnt by children as an alternative to the *gojūon*, but also used in assigning *kana* to successive articles or clauses in an official document. The order of words in dictionaries and directories is, however, based on the *gojūon*.

The ordering of *kanji* is much more complicated. The starting point is that any *kanji* has a fixed number of strokes,[59] which are written in an order governed by rules which children must learn at primary school. Every *kanji* is taught separately, and the children in class, following their teacher, will draw the strokes in the air, in the correct order, and counting '*ichi, ni, san* ...' so that in the end a cardinal number (that of all the strokes comprised in the character) is reached by an ordinal process. Since a child will go through this process thousands of times in six years of primary school, it gains an unusually thorough experience of simple counting.

The *kanji*, as a class of signs, has no inherent order, any more than the letters of an alphabet. There is however the important difference that whereas the alphabet is a closed set, with a small number of members which can be learnt without paying much regard to their relative importance, *kanji* must be treated as an open-ended set for which no definitive, comprehensive order can be established. Some principle of order is indispensable, however, whether for determining the *kanji* syllabus in primary school or the way in which a dictionary must be organised.

As for the syllabus, this has been established, since 1949, according to a list now known as the *jōyō kanji*.[60] This establishes something less than 2,000 *kanji* as basic, of which something under a half must be learnt in prescribed stages corresponding to the six years of primary school, leaving the rest to be picked up in the first three years of secondary school. The *jōyō kanji*, generally still referred to by the earlier name *tōyō kanji*, are a popular institution even for those Japanese who completed their school education before the lists were introduced. They represent a true case of the invention of tradition, in the sense of one 'actually invented, constructed and formally instituted' (Hobsbawm and Ranger 1983: 1). In the context of numbers they ram home the lesson that the mastery of any collection of objects is best achieved by first establishing them in a fixed numerical order. When this relates to time, it defines the necessary strategy for achieving the desired result – in this case the mastery of written Japanese.

For the organisation of dictionaries quite different rules are applied, which take no account of how important a given *kanji* may be in actual usage. A modern *kanji* dictionary almost always lists the words contained in it according to the *gojūon* order of the *on* readings.[61] At first sight this assumes that these are known to the user, but this is in fact not necessary, since all the *kanji* dealt with are listed at the end of the dictionary according to a system based primarily, if not quite entirely, on their actual visible form.

The traditional system depends on every *kanji* having two components, one *radical* and the other *phonetic.* There are over 200 recognised radicals, of which less than a half occur at all commonly. The principle is that the radicals divide the *kanji* into different categories, so that, for example, those incorporating the 'wood' radical, 木, will indicate some sort of tree. It is probably true that over the broad spectrum of written Japanese the rule is honoured more in the breach than in the observance, but it is sometimes a useful if never a reliable guide to meaning. In any case most Japanese grow up to believe in it.

The first rule, then, in finding the meaning of an unknown *kanji* is to identify its radical (which occasionally will correspond to the whole). The number of strokes in the radical will determine its place in the list in the dictionary. The number of strokes left over define the phonetic component, which may actually indicate the correct pronunciation of the *on* reading.[62] This number then indicates the place of the *kanji* under the relevant radical.

The working of the rules is best illustrated by an example which actually adheres to them. Taking the character 時, which has already been introduced, this is a familiar character learnt in the second year of primary school according to the *jōyō kanji* programme. Even a 7-year-old child will have learnt that 日 is the only possible radical, since the remaining part, 寺, is not on the list. He will know also that 日 is the 'sun' or 'day' radical, and as such has the number 72. The part left over, the phonetic component, has six strokes, so by turning to the *kanji* listed under the heading 72 and the sub-heading 6, the child will soon find it, for there will only be two or three others meeting this exact specification. Once the *kanji* has been found, the number printed next to it, say 232, will give the page of the dictionary entry.

The final result is that 時 has the *on* reading, *ji,* and the *kun* reading, *toki,* and the explanations given will show that it means the Japanese equivalents of 'time' (as has already been noted earlier in this chapter). The precocious child would note that the phonetic component, 寺, also has the *on* reading, *ji,*[63] and means 'temple' in such compounds as *kinkaku-ji,* the well-known golden temple of Kyoto.

The lesson is that mastery of *kanji* can only be achieved by applying, according to whether the perspective is that of the teacher or the pupil, an appropriate numerical procedure. Learning *kanji* involves, inescapably, learning to work with numbers in an extremely practical way. The procedures, if involved, have no mathematical depth. This is why they can be mastered by such young children. The problem lies not so much in mastering the numerical procedures, but in keeping abreast of the semantics of what they govern, that is the unending series of ever more abstruse *kanji.*

So long as *kanji* continue to be used the whole procedure outlined above will be an inescapable part of growing up, unless, at least, a radically different syllabus is adopted. The basic *tōyō kanji* syllabus is now part of the life-time experience of all Japanese under 50, and so pervasive is it that it has penetrated the consciousness of almost all those over that age. If the explanations given above are involved, they are still essential, for in Japan, as Unger has noted, 'all literacy, and hence all education, is grounded in *kanji*' (Unger 1987: 84).

# 3 Alternative number systems

## THE TYPOLOGY OF NUMBER SYSTEMS

If, intuitively, we think of numbers in terms of the infinite series of natural numbers, 1, 2, 3, 4, ..., we are then equally likely to take for granted that they are correctly represented by a place-value system, based on the number 10, such as that of the familiar, so-called Arabic numerals. I have shown in the first two chapters that this understanding of numbers is not necessarily essential, but largely the recent product of modern western culture, although its roots are certainly very much older. Even the functional supremacy claimed for the place-value system is based on the assumption that the primary use of numbers is defined by contexts which require them to be subjected to the four elementary arithmetical operations of addition, subtraction, multiplication and division. Modern Japan also takes for granted that these four operations, known as *shisoku*, define the primary form of numbers, which in the Japanese case also includes their representation by the beads of an abacus.

There is no logical necessity for any of the intuitive assumptions outlined in the previous paragraph. In the end it is the first of these assumptions, that the series of numbers is infinite, which is the most questionable. True, this proposition may be *inherent* in the logical definition of cardinal numbers, but then what is the necessity for allowing such numbers any logical priority? Are they an essential part of every numerical culture, and even if they are, must every such culture include the logical conclusions of western philosophy?[1] It may be that modern Japan, as much as the western world, takes for granted that numbers are in some way *made* for arithmetic, but in Japan, in particular, there are any number of number systems which are so ill-suited for arithmetic that no attempt is made to subject them to arithmetical operations. Once a number system does not have to satisfy this requirement, there is no longer any need for a place-value system, or a single numerical base. This presents the problem of conceiving of numbers in a form deprived of attributes which are, intuitively at least, implicit in their most common form of representation.[2] This is, of course, the series 1, 2, 3, ....

The problem is essentially cultural, or perhaps better, cognitive. If one transfers it out of the domain of a culture dominated by the use of numbers for measuring and bookkeeping, one experiences the sort of freedom which Lewis Carroll bestows upon Humpty Dumpty (see Gardner 1970: 269):

> 'When *I* use a word,' Humpty Dumpty said, in rather a scornful tone, 'it means just what I choose it to mean – neither more nor less.'
> 'The question is,' said Alice, 'whether you *can* make words mean so many different things.'
> 'The question is,' said Humpty Dumpty, 'which is to be the master – that's all.'[3]

In the context of alternative number systems, this leaves it open to develop and maintain just as many such systems as the local culture requires. What is more, the meaning of any such system may be adapted to suit any given context for which it is appropriate. To avoid confusion, the *number words* may also depend on context, so that a distinction is made between the words for the four cardinal points and the four suits in a pack of cards,[4] although the numerical systems they define have important common attributes. In the present chapter I look at five alternative numbers systems, all of which have a number of different uses within Japanese culture. These do not exhaust all the possibilities. Two domains, song and games, are characterised by the use of number systems specific to the contexts which arise. The first of these is dealt with at the end of the present chapter, the second is the subject matter of chapter 8.

## YIN–YANG, MANDALA AND ONYŌDŌ

The three systems which I describe in this section are all binary. *Yin–yang* is the basic symbolism of such systems in the traditional culture. The meaning is simply that the two component elements, known in Japanese as *in* and *yō*,[5] are, according to circumstances, both complementary and in opposition to each other. *In–yō* is essentially a principle of order born out of a time when the universe was in a state of primordial chaos. At this first stage, the universe may be seen as simply *uchū*, or a cosmos with ten thousand (*ban*) different images (*shō*), represented by the expression *shinra-banshō*, which evokes all the different forms of life to be found in the luxurious growth (*shinra*) of a forest. But *uchū* is also *tenchi*, that is heaven and earth, and *in–yō* is what separates the one domain from the other. In this process *yō* is *ten*, and *in*, *chi*. The relationship between *in* and *yō* is then to be found in three representations of *tenchi*. The first of these, *tenchi dōkon*, expresses the common root, *dōkon*, of heaven and earth. The second representation is *tenchi ōrai*, expressing the interaction between the two, and the third is *tenchi kōgō*, the union between them. *Kōgō* is strictly sexual union, so that it continually gives rise to new creation; in the end, therefore, something close to the primordial *banshō* is restored. Starting at the origins of the universe, *in* and *yō* provide the basis for explaining and interpreting everything man experiences in his daily life, at the same time pointing the way to future action.

It is hardly surprising that the relationship between *in* and *yō* has any number of metaphorical representations, so that *in*, the negative principle, connotes the female, darkness (and by implication the north side of a mountain or the south side of a river), secrecy, the moon, where *yō*, the positive principle, connotes the male, light (and so the south side of a mountain and the north side of a river), openness and the sun. Although *in* and *yō* lend themselves to be used metaphorically in any context requiring a binary opposition, such as between darkness and light, they can equally represent the two digits, which may be taken to be 0 and 1, by definition required of any binary system of numeration. In principle, such a system can represent the infinite series of natural numbers just as effectively as the decimal system of Arabic numerals, and indeed the modern computer is continually translating from one system to the other. This, however, is of little importance in the present context. More significant is the possibility of establishing a succession of closed systems, generally by defining an upper limit, $n$, to the number of digits which may occur in the numbers represented.

Any such system, defined by a constant, $k$, will then contain $2^k$ numbers, and being a closed system there is no essential requirement for one order to be preferred over all others. In other words, there is no inherent virtue in any one rule for ordering the members of a closed system. This is in pronounced contrast to the rule required to express the infinite series of natural numbers in binary form, which, in any elementary case,[6] must be that a lower number always precedes any higher number. With Arabic numerals this establishes the familiar counting order, 1, 2, 3, 4, 5 ...; the binary case is more opaque, 1, 10, 11, 100, 101 ..., but essentially the same. The price paid for abandoning this rule for any given system is that it excludes any possible application requiring standard arithmetical operations. This point has already been made.

There are two possible consequences. The first is that the numbers comprised in the closed system will be used exclusively for defining relations, and attributes, external to it. This, as I show in chapter 6, is the basis of all numerically based fortune-telling, such as, for example, horoscopes. This is the point made by Humpty Dumpty (who may be seen as a putative fortune-teller). It explains why so many such systems, such as that defined by the animals in the zodiac, do not appear to use numbers at all. The second possibility is that the numbers comprised in the closed system will be capable of being combined by means of prescribed operations, but that these operations will be defined *ad hoc* and will not be those of standard arithmetic. This is true of any number of games which require, in practice, the players to master the operations which define them. In chapter 10, I describe a number of special Japanese cases.

The point made in the previous paragraph is entirely general, and its validity is not confined to binary systems. None the less, in this section, I confine its application to such systems, if only because these are so well suited to illustrate it. In the following section I introduce the system based on the five elements, which will prove to be an extension of the same basic principles.

The *yin–yang* opposition provides the basis for the eight *trigrams*, known in Japanese as *hakke* ( 八卦 ). The trigram consists of three lines, each of which may be broken (*in*) or unbroken (*yō*), so that there are the eight possible combinations given in table 3.1. These in turn combine, in pairs, to constitute the sixty-four hexagrams which are the basis of the Japanese art of divination known as 'onyōdō' ( 陰陽道 ). This is simply *inyō*, in the alternative form, *onyō*, followed by 道 meaning 'way' – the familiar Chinese *tao*.[7] The important point is that the sixty-four hexagrams, each comprising six broken or unbroken lines, corresponding to *in* and *yō* (Ronan and Needham 1978: 161), are in fact no more than an extension of the binary system of numeration. There was, however, 'no Chinese thought of binary arithmetic or even any realisation that such counting could exist' (ibid.: 189),[8] and certainly the Japanese, in adopting the lore of *yin* and *yang* under the name *onyōdō*,[9] did no better. The sixty-four hexagrams are a closed system, which each combination having a recognised meaning according to the context in which it occurs.

*Table 3.1* Trigrams

| Hakke | kanji | name | meaning | word |
|-------|-------|------|---------|------|
| ☰ | 乾 | *ken* | heaven | *ten* |
| ☷ | 坤 | *kon* | earth | *chi* |
| ☳ | 震 | *shin* | thunder | *rai* |
| ☴ | 巽 | *son* | wind | *fū* |
| ☵ | 坎 | *kan* | water | *sui* |
| ☲ | 離 | *ri* | fire | *ka* |
| ☶ | 艮 | *gon* | mountain | *san* |
| ☱ | 兌 | *da* | lake | *taku* |

Numerical factors played, at best, only a very minor part. The fact that there was no one order for the hexagrams almost excludes the possibility of their being seen as isomorphic to a recognised *series* of numbers. Whatever algorithm may have determined the order adopted in any given context,[10] the only governing principle was the alternation of *yin* and *yang*.

This is generally the form of fortune-telling known in Japan as *ekikyō*, and in China as *I Ching* (under which name it has become popular in the west). The Japanese version of this system I describe in chapter 6.

The *mandala* (or in the standard Japanese form, *mandara*) is a binary system of a quite different kind.[11] The *mandala* is an ideal topological construction,[12] comprising a core (*manda*) and a container (*-la*). The core consists of a single unit, whereas the number of units comprising the container is always a power of 2. This, the *simple* model, is built up out of $2^n + 1$ units,[13] which means that its binary representation is always in the form of two '1's, separated by a series of '0's.[14] The complex model has a series of containers, each with a number of units counted in increasing powers of 2, so that any such model can be defined by a binary code, representing, at the same time, the total number of units it contains. Examples of both the simple and the complex models are given in figure 3.1. In mathematical terms there is in principle a perfect one-to-one correspondence between the series of binary odd numbers,[15] and the class of all possible mandalas. This means that a *mandala* can

*Figure 3.1* Simple and complex mandala

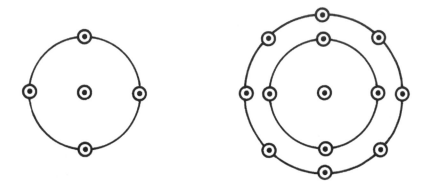

always be designed to represent any odd number,[16] save one and possibly three.[17]

In practice, although any odd number provides a blueprint for the design of a *mandala*, the general function of such a number is to represent a particular mandala, whose actual design was determined by other factors. In other words, the design of the mandala is seen as logically prior to the number which represents it. The point is confusing, since there is a one-to-one correspondence between the number and the design, and this is not lost upon the cultures in which the *mandala* occurs.[18] Although in chapter 8 I examine a number of instances, known from Japanese culture, of the *mandala*, none of these is as important as say, Angkor Wat, in Cambodia, or Borobudur in Java.

## GOGYŌ OR THE FIVE ELEMENTS

The five elements of ancient Chinese science, wood, fire, earth, metal and water, are known in Japan under the name of *gogyō*, although the way in which they are used, even today, is largely of Chinese origin. The actual order in which the *gogyō* occur may vary according to context, which also determines their metaphorical, as opposed to their literal meaning. This is important, for the *gogyō* relate to any number of sets containing five members,[19] and for some of these they also provide the names. In particular, the five planets known to the ancient world, were named in China after the five elements, and these names were adopted in Japan. It follows that Jupiter is *mokusei* (wood star), Mars, *kasei* (fire star), Saturn, *dosei* (earth star),[20] Venus, *kinsei* (metal star) and Mercury, *suisei* (water star). The five planets are seen as born of the union (*kōgō*) between sun and moon, so that *in* and *yō* combine to produce the *gogyō*.[21] The birth of the planets, if but one representation of this mystical union, may still be taken as the archetype.

There is also an essentially mathematical relationship between *inyō* and *gogyō*. This is based on the principle that both systems are essentially cyclical. With *in* and *yō* this causes no difficulty: cyclical order means, in this case, no more than continuous alternation, such as occurs between day and night or winter and summer. With *gogyō* the position is more complicated, but the basic mathematical principle is that there are only two essentially different orders, which are complementary to each other. This is illustrated by figure 3.2, in which the essential order can be stated in terms of which elements are adjacent to each other. (This means that the order of the mirror image is taken to be the same.) Taking the order in figure 3.2a as the starting point, the only essentially different order is one in which elements previously adjacent to each other are all separated. To see what this involves, one can start with any one of the five elements, say *water*. The required transformation allows it to remain adjacent to neither metal nor wood, so that these elements must be supplanted by earth and fire. Earth, in turn, may no longer be adjacent to metal or fire, so that with water on one side it can only have wood on the other. So also, fire, with water on one side, must have metal on the other. This produces the order given in figure 3.2b. Repeating the process leads back to figure 3.2a. If the vertices in figure 3.2a are joined in the cyclical order of figure 3.2b, the result is the five-pointed star in figure 3.2c. The star, and the pentagon defined by its vertices, then represent the two possible configurations.

This transformation is not possible with less than five elements, and with more than five elements the alternation between the two figures is, mathematically, no longer essential.[22] The Japanese recognised the alternation and gave a positive value to one configuration, which was designated *sōshō* to mean 'harmony', and a negative value to the other, then designated *sōkoku*, to mean 'antagonism'.[23] At the same time, the relationships were given direction, as indicated in figure 3.2d. Where it is a case of *sōshō*, the direction of the arrows indicate that one element is born out of another. How this is supposed to happen is part of the tradition. Wood, for instance, gives birth to fire, since fire is made by rubbing two pieces of wood

*Figure 3.2* Gogyō

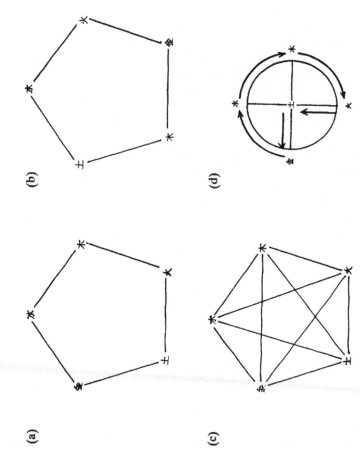

(a)

(b)

(c)

(d)

Key to kanji: Earth 土, Metal 金, Water 水, Wood 木, Fire 火.

together. In the case of *sōkoku*, the direction of the arrows indicates that one element destroys the other, so that wood destroys earth, since the roots of trees break up the ground. At the same time every force implies its opposite, so that every case of *sōshō* has an element of *sōkoku* and vice versa. If there is always a cloud on the horizon, every cloud still has its silver lining. If the roots of trees destroy the earth, they can also bind it together and prevent landslides.

Given that *gogyō* was a multivalent system, capable of indefinite extension, the ambiguity of *sōshō* and *sōkoku* made certain that in any context practitioners could always be wise after the event if things went wrong. *Gogyō*, as much as any other of the alternate number systems used in Japan (and generally imported from China), was designed to be a part of applied mathematics, so the applications could prove to be counter-productive.

Figure 3.2d also places one element, *do* (earth), at the centre, which was essential for allowing the lore of *gogyō* to adapt and incorporate systems containing four members, such as the cardinal points or the seasons of the year.[24] The cycle of *in* and *yō* can also be made to include the five elements, by taking earth to be the perfect balance between the two, fire as pure *yō*, water as pure *in*, wood as *in* in *yō*, and metal as *yō* in *in*.[25] This makes it possible in any application combining *in–yō* and *gogyō* to use one system to interpret results derived from the other.[26]

## KANSHI

The oldest of all the alternative number systems used in Japan is that based on a series of twenty-two characters or cyclical signs which are to be found among the most ancient Chinese written texts. These twenty-two *kanji*, as given in figure 3.3, are familiar to the Japanese in any number of contexts, but their most common use is in the almanacs which, every year, are sold by the million in every possible retail outlet.

The twenty-two signs divide into two groups, one containing the ten *kan*, or 'heavenly stems', and the other, the twelve *shi*, or 'earthly branches'.[27] This is another case of a dichotomy between heaven and earth, such as also occurs in the lore of the five elements. If the original meanings of the signs is lost in antiquity, there is little doubt about how they are now interpreted. The ten *kan* are known to the Japanese in two forms, both given in figure 3.3. The first is a single *on* reading, while the second is a reading based on combining *kun* readings of the five elements, in the order wood, fire, earth, metal, water,

*Figure 3.3* **Kanshi**

*Note:* The first of the two names for each of the combinations is a combination of the *on* readings of the two symbols. Note that the combination of *ko* and *shi* (number 1) can be pronounced *kasshi* and that *otsu* becomes *itsu* in combination and undergoes other sound changes. The second of the two names combines an alternate name of the stem with the animal name for the branch.

| No. | Kanji | Name 1 | Name 2 |
|---|---|---|---|
| ① | 甲子 | kōshi (kasshi) | kinoe ne |
| ② | 乙丑 | itchū | kinoto ushi |
| ③ | 丙寅 | heiin | hinoe tora |
| ④ | 丁卯 | teibō | hinoto u |
| ⑤ | 戊辰 | boshin | tsuchinoe tatsu |
| ⑥ | 己巳 | kishi | tsuchinoto mi |
| ⑦ | 庚午 | kōgo | kanoe uma |
| ⑧ | 辛未 | shimbi | kanoto hitsuji |
| ⑨ | 壬申 | jinshin | mizunoe saru |
| ⑩ | 癸酉 | kiyū | mizunoto tori |
| ⑪ | 甲戌 | kōjutsu | kinoe inu |
| ⑫ | 乙亥 | itsugai | kinoto i |
| ⑬ | 丙子 | heishi | hinoe ne |
| ⑭ | 丁丑 | teichū | hinoto ushi |
| ⑮ | 戊寅 | boin | tsuchinoe tora |
| ⑯ | 己卯 | kibō | tsuchinoto u |
| ⑰ | 庚辰 | kōshin | kanoe tatsu |
| ⑱ | 辛巳 | shinshi | kanoto mi |
| ⑲ | 壬午 | jingo | mizunoe uma |
| ⑳ | 癸未 | kibi | mizunoto hitsuji |
| ㉑ | 甲申 | kōshin | kinoe saru |
| ㉒ | 乙酉 | itsuyū | kinoto tori |
| ㉓ | 丙戌 | heijutsu | hinoe inu |
| ㉔ | 丁亥 | teigai | hinoto i |
| ㉕ | 戊子 | boshi | tsuchinoe ne |
| ㉖ | 己丑 | kichū | tsuchinoto ushi |
| ㉗ | 庚寅 | kōin | kanoe tora |
| ㉘ | 辛卯 | shimbō | kanoto u |
| ㉙ | 壬辰 | jinshin | mizunoe tatsu |
| ㉚ | 癸巳 | lishi | mizunoto mi |
| ㉛ | 甲午 | kōgo | kinoe uma |
| ㉜ | 乙未 | itsubi | kinoto hitsuji |
| ㉝ | 丙申 | heishin | hinoe saru |
| ㉞ | 丁酉 | teiyū | hinoto tori |
| ㉟ | 戊戌 | bojutsu | tsuchinoe inu |
| ㊱ | 己亥 | kigai | tsuchinoto i |
| ㊲ | 庚子 | kōshi | kanoe ne |
| ㊳ | 辛丑 | shinchū | kanoto ushi |
| ㊴ | 壬寅 | jin'in | mizunoe tora |
| ㊵ | 癸卯 | kibō | mizunoto u |
| ㊶ | 甲辰 | kōshin | kinoe tatsu |
| ㊷ | 乙巳 | isshi | kinoto mi |
| ㊸ | 丙午 | heigo | hinoe uma |
| ㊹ | 丁未 | teibi | hinoto hitsuji |
| ㊺ | 戊申 | boshin | tsuchinoe saru |
| ㊻ | 己酉 | kiyū | tsuchinoto tori |
| ㊼ | 庚戌 | kōjutsu | kanoe inu |
| ㊽ | 辛亥 | shingai | kanoto i |
| ㊾ | 壬子 | jinshi | mizunoe ne |
| ㊿ | 癸丑 | kichū | mizunoto ushi |
| 51 | 甲寅 | kōin | kinoe tora |
| 52 | 乙卯 | itsubō | kinoto u |
| 53 | 丙辰 | heishin | hinoe tatsu |
| 54 | 丁巳 | teishi | hinoto mi |
| 55 | 戊午 | bogo | tsuchinoe uma |
| 56 | 己未 | kibi | tsuchinoto hitsuji |
| 57 | 庚申 | kōshin | kanoe saru |
| 58 | 辛酉 | shin'yū | kanoto tori |
| 59 | 壬戌 | jinjutsu | mizunoe inu |
| 60 | 癸亥 | kigai | mizunoto i |

with the *kanji* for older and younger brother represented by the deviant readings *e* and *to*. In the lore of *yin–yang*, the former corresponds to *yō* and the latter to *in* (Yoshino 1983: 38). Although, in this way, the ten *kan* are related to both *gogyō* and *onyōdō* (which establishes their place in a unified theory of the cosmos), this relationship plays little part in their use in conjunction with the twelve *shi*.

The twelve *shi* represent the twelve animals of the Chinese zodiac, and although each has its own recognised characteristics,[28] once again the way in which they combine with the ten *kan* is what counts in the numerical system. The number of combinations occurring is sixty, the lowest common multiple of ten and twelve. The order is that given in table 3.2. The combined *kanshi* system is thus appropriate for measuring any cycle containing sixty members, and the system itself has defined any number of such cycles, particularly relating to the passage of time, and its use in horoscopes. These I discuss in chapters 6 and 7.

## SONG

The Japanese *uta* (歌) means both 'song' and 'verse'. Although, for reasons given in the Preface, nothing can be said about traditional Japanese music, the best-known verse forms are defined numerically. These are the 17 syllable *haiku*, and the 31-syllable *tanka*, which is also known as *waka*.[29] The former has three lines, with 5–7–5 syllables, and the latter, five, with 5–7–5–7–7 syllables. Originally the *tanka* was the shortest possible form of a *chōka* or 'long song'[30] which had an indefinite number of 5–7 pairs, ending always with an additional seven-syllable line. At the present day the *tanka* and the *haiku* have become the standard verse form, while *chōka* are hardly composed at all. The *tanka* retains its ancient form, whereas the *haiku* was only established as an independent poetic form at the end of the nineteenth century.

Compositions in both forms are now produced by both professionals and amateurs. The latter, in particular, have a chance of competing in the annual poetry-reading, which is part of the new year's celebrations of the imperial household. Members of the imperial family are ardent poets, and the Emperor Meiji is recorded as having composed more than 10,000 *tanka*. Many Japanese spend the weeks before the end of the year composing *haiku* for their friends, which are then transcribed in calligraphy on the new year cards which can be bought at any post-office. These are sold by the million; each card has a separate number printed on it, which the

*Table 3.2* Jikkan jūnishi

(a) The ten stems or trunks (*jikkan*)

(b) How the five elements are combined with *yin* and *yang* to give alternate names to the ten stems

(c) The twelve branches (*jūnishi*)

(a)

| Chinese character | 甲 | 乙 | 丙 | 丁 | 戊 | 己 | 庚 | 辛 | 壬 | 癸 |
|---|---|---|---|---|---|---|---|---|---|---|
| On reading | kō | otsu | hei | tei | bo | ki | kō | shin | jin | ki |

(b)

| Five elements | 木 *ki* (wood) | | 火 *hi* (fire) | | 土 *tsuchi* (earth) | | 金 *kane* (metal) | | 水 *mizu* (water) | |
|---|---|---|---|---|---|---|---|---|---|---|
| E and to (yang and yin) | e (elder) | to (younger) | e | to | e | to | e | to | e | to |
| Name | kinoe | kinoto | hinoe | hinoto | tsuchinoe | tsuchinoto | kanoe | kanoto | mizunoe | mizunoto |

(c)

| Chinese character | 子 | 丑 | 寅 | 卯 | 辰 | 巳 | 午 | 未 | 申 | 酉 | 戌 | 亥 |
|---|---|---|---|---|---|---|---|---|---|---|---|---|
| On reading | shi | chū | in | bō | shin | shi | go | bi | shin | yū | jutsu | gai |
| Animal name | ne | ushi | tora | u | tatsu | mi | uma | hitsuji | saru | tori | inu | i |
| English | rat | ox | tiger | hare or rabbit | dragon | snake | horse | ram or sheep | monkey | rooster | dog | boar |

*Note.* The "no" of these combinations is the genetive particle. "*Kane*" becomes "*ka*" in its two combinations.

recipient may then treat as a ticket in a national lottery,[31] for which the draw occurs early in the new year.

The actual verse form is quite undemanding, so long as the right number of syllables occurs in each line. Two contrasting but related themes are often expected, so the writer must be extremely economical with words. Strict adherence to grammar is not required. Motoori Norinaga, the great eighteenth-century literary scholar, insisted that only autochthonous words (i.e. *kun* readings of *kanji*) be used, but only purists would adhere to this rule today. In popular culture, verse, or *uta*, is more or less synonymous with the 17- and 31-syllable forms of the *haiku* and the *tanka*, which are what the two numbers immediately connote for any Japanese.

# 4 The culture of numbers

## CULTURE AND LANGUAGE

In the two previous chapters I outlined the forms in which numbers are known to the Japanese. These forms are essentially linguistic, although in many cases the numerical content is not explicit. These latter cases, by and large, are those dealt with in chapter 3, which is essentially a lexicon for the systems examined, their actual application being the subject matter of later chapters. The present chapter deals with the opposite case, defined by a very large number of instances in which the numerical content is perfectly explicit but its meaning remains allusive. Paradoxically, this one explicit element – that is the actual value of the number used – may be of no importance, so that another number could be substituted without any change in meaning. In such a case, the common reason is that the number actually occurring does no more than represent some general class of which it is a member. An example will make this point clear. In the Dutch version of the English proverb, 'A bird in the hand is worth two in the bush', 'two' is replaced by 'ten',[1] but the meaning, plainly is unchanged. All that is needed, in either case, is that the second number be greater than the first (or at least within reasonable bounds, for a very large discrepancy would certainly change the meaning).[2] This example is significant for the existence of another saying, 'half a loaf is better than no bread', with much the same meaning.

The cultural significance of such use of numbers is that an abstract principle is reduced to an instance of its application, which is then accepted as a shorthand statement of the principle itself. This is essentially an instance of *numerical metonymy*, a form of number use prolific in Japanese culture. Expressed algebraically, the principle, in the case chosen for illustration, is that if $a > b$, circumstance may require that $b$ is still to be preferred to $a$. A difficult problem of probability is simply by-passed; if one is content with the bird in the hand, one is free from any need to calculate the chances of catching both the birds in the bush (or, in Holland, all ten in the air).[3]

In the cases in which the magnitude of the number appears to be significant, this may still be defined *post hoc*. We are used to there being *three wise men*, but the New Testament does not mention any number,[4] and none is needed, even though later tradition has actually provided the three wise men with names.[5] So also, in Japan, the well-known story of the forty-seven *rōnin*, could do equally with any other number of the same order, although few Japanese would readily accept this. Even where a number explicitly counts the members of a class, it may still, in principle, be adventitious. Would Christianity have any difficulty in finding an eighth deadly sin or in reducing the number to six? But principle is essentially culture-bound, and a Japanese would no doubt find it absurd to consider altering the number of the *five essential relationships* (Gojō) in Confucianism: they are seen as defining the whole structure of its authority (Moeran 1985: 94f).

It remains true, none the less, that in relation to the number of possible abstract principles the number of stock phrases, aphorisms and proverbs containing numbers is very large, particularly in such a rich numerical tradition as that of Japan. The only way to deal with this wealth of material is to select certain typical cases for illustrating the different types of number use. This is not only the scheme of the present chapter, but also that of chapter 5, dealing with the special case of Japanese names. The starting point, therefore, for each of the following sections, will be the Japanese lexicon, which contains countless stock phrases incorporating numbers.

The analysis will then proceed to the occurrence of numbers in proverbs and aphorisms. Once again, the question will constantly arise as to the generality of the ideas and principles expressed. The richness of the material will be apparent at every stage, leaving one to ask how much of it is distinctively Japanese.

The material dealt with in this way can be divided into a number of distinct cases, each of which has its own section. At the very first level a distinction is made between expressions containing one and those containing more than one number. In the former case the meaning of the phrase can only depend on some attribute of the number used, while in the latter it is the relationship between the numbers which is important.

The instance of the single number is in turn divided into three distinctive cases, each with their own section. The first of these

sections is devoted simply to the number 1, which is very much a case
of its own. The following section then deals with numbers within the
range of normal counting, so that many of the instances cited will deal
with numbers under 10. (This section will conclude with a small
number of cases of numbers with special connotations derived from
their written forms, a phenomenon which is entirely dependent upon
the Japanese use of *kanji.*) Very large numbers are then a quite
separate case, with their own section.

This leaves the case of number combinations to the final section:
this deals with a number of separate instances, each with their own
governing principle. This in turn is determined by the recognised
relationship between the numbers used, of which possible instances
are given by the contrast, in English, between such expressions as 'at
sixes and sevens' and 'one in a hundred'. In this section, in particular,
I seek to establish, in the Japanese context, a sort of *algebra of
metonymy.*

## THE UNIQUENESS OF 1

The number 1, the obvious starting point, proves to be a very special
case.[6] 'One is one and all alone, and ever more shall be so',[7] so the
Japanese *hitori*[8] *de,* literally 'by one man' simply means 'alone'. This
meaning is made more specific in a phrase such as *ITte ni* which
almost literally translates the English 'single-handed'. If there is but
one, then that one is also the whole, as is implicit in such English
words as the *uni*verse. In Japanese therefore, *ISsai* means 'everything, the absolute', *IPpan [no],* 'general' or 'ordinary', and *ITtei,*
'uniformity, certainty', with the principle being extended to the word
*ICHImi,* for a 'gang of fellow conspirators' – so that in Japan, apparently, there is honour among thieves.

Not only is one all-embracing, it is also the element which cannot
be broken down any further, as is implicit in the English, 'one
moment, please' or its Japanese equivalent *IKkoku hayaku,*[9] which
combines the moment of time, with the concept of speed. A more
evocative variant is *ISshunkan,* literally 'in the interval of one wink',
or in idiomatic English, 'in the twinkling of an eye'. In such a context
*one* can be used redundantly, so that *ikkyo ichidō* means, in a self-
deprecating sense, 'what little one does'.

In any ordered series *one* is far from negligible; it is the first
member of the series, so that to such English expressions as 'A1'
correspond the Japanese *ITtō,* or simply *ICHIban,* the normal word
for 'most'. In another combination *ISshin* connotes 'brand new', a

complete change of model. *Ichi* also connotes 'basic' or 'fundamental', so that *ICHIgi* is a literal translation of the English expression,
'first principles'.

What has been given so far is no more than a short, representative
list of dictionary expressions, but this is really no more than a
beginning when it comes to exploring all the recognised connotations
of the number 1 in Japan. Mori (1980) alone lists several hundred,
and many of these are multiple entries. An example is *HITOkuchi*,
literally 'one mouth', for which Mori (ibid.: 43–4) lists nine separate
meanings, of which two are specific to a single prefecture. All have in
common the idea of the smallest unit in which something can occur,
as in Toyama prefecture, a single gust of wind. The same is true of the
three extensions listed, *-akinai*, *-jōruri* and *-mono*, which, in the
different contexts of commerce, drama and consumption, define a
meaning in terms of a single transaction or event, so that *HITO-
kuchi-mono* is an hors d'oeuvre, or anything else which can be taken
as a single mouthful. This principle of reduction is carried to its
extreme in the name of the god, HITOkoto Nushi no Kami, whose
seat is Mount Kazuraki on the frontier between Osaka and Nara.
According to tradition,[10] the Emperor Yūryakyu (AD456–79)
disputed with the god whether a single principle could embrace both
good and evil, a theological point known also outside Japan.[11]

Some uses of the number 1 are beautifully evocative. An example
is *ITtenkō* (ibid.: 38), which can mean a 'pomegranate' (instead of
the normal *zakuro*), but also a particular red wild flower reminiscent
of the European poppy. This leads to a popular metaphorical use in
which *ITtenkō* is transformed into *kōITten*, an expression for the
only woman in a men's party. This, once again, is the connotation of
uniqueness.

There are also any number of proverbs, or *kotowaza*, containing
the number 1. These not only reflect the different aspects of the
number already considered, but also express ideas occurring in
proverbs in other parts of the world. (This will prove to be true of
many proverbs not including the number 1.) 'One swallow does not
make a summer' is almost literally translated by, *ICHIen natsu wo
nasazu*, but consider then, *ICHIyō ochite tenka no aki wo shiru*, or
'with the first fallen leaf[12] one knows that autumn has come'.[13] This is
an instance of another common case, that is, of two proverbs
expressing contradictory ideas.[14]

The idea of *one* representing the smallest unit that can occur is
reflected in the many Japanese proverbs based on *ISsun*, roughly
equivalent to 'one inch'. In general, the proverbial meaning seems to

be that just because something is small, it is not to be despised; indeed, if anything, the message is that a small difference can be crucial.[15] *ISsun saki wa yami*, meaning literally 'one inch forward is darkness', recalls the familiar English idiom, 'a step in the dark'. The proverb is actually understood to mean that the future is always an unknown quantity. So also, *ISsun no kōin karonzu bekarazu* is equivalent to, 'a lost moment can never be retrieved'. The potential of something small to become immeasurably greater reflected in the English, 'from a single acorn comes a great oak', is to be found also in the Japanese, *ISsun no mushi nimo gobu no tamashii*, literally, 'a one-inch insect is still five parts spirit'. Once again the metaphor is not consistently applied, so that the whole tenor of the Japanese, *ISsun kararuru mo NIsun kararuru mo onaji koto*, is that a one-inch cut and a two-inch cut cause the same pain.

Two of the examples given in the previous paragraph contrast the number 1, with another number. Such contrasts, not necessarily involving the number 1 are extremely common in Japanese proverbs and aphorisms – so much so that they are dealt with below in a separate section. The different contexts in which *one* occurs all tend to emphasise its uniqueness: its referents are not those of other numbers, so that if some other number were substituted, the expression would lose its meaning.

## COUNTING NUMBERS

In any numerical culture in which language offers the means of expressing very large numbers, beyond the range of counting, there is an implicit distinction between the contexts in which such numbers occur and those in which smaller numbers, within the range of counting, are to be found. In the domain of the present chapter, the contexts of numbers not greater than 10, and those of numbers not less than 1,000, are quite different. An English example will help make the point clear. The meaning of the proverb, 'One picture is worth ten thousand words', depends on the fact that 10,000 is a very large number: this makes the point that a visual image is far more economical as a means of communication than any spoken utterance.[16] Drop the *thousand* and the meaning changes entirely, for ten words require far less effort than even the simplest sketch: the proverb is not even plausible. One can turn the argument on its head by looking at a proverb like, 'A stitch in time saves nine'; in this case substituting 900 for 9 would destroy the whole balance of the saying.

The distinction proves to be equally important in Japanese. The problem, once again, is the enormous wealth of examples which illustrate it. In the lexicon the lowest numbers, and 2 in particular, occur in combinations analogous to many words occurring in English. In these cases the Japanese version tends to be more transparent, so that *FUTAgo* for 'twins' means literally 'two children', or *FUTA-gokoro*[17] for 'duplicity', 'two minds'. Not surprisingly the distinction between *yin* and *yang* or in Japanese, *in* and *yō*, is summed up in the expression *NIki*, meaning 'two winds'.[18] More generally, the number 2 suggests the contrast between any two opposites, such as black and white;[19] this is summed up in the general abstract principle of *NIge-datsu*, contrasting the two different ways to salvation recognised by Buddhism. The contrast is expressed in many different ways. It is that between the active (*ui*) and passive (*mui*) principle, or between inborn (*sei*) and acquired (*toku*) characteristics, or heart (*shin*) and mind (*kei*), which one finds, of course, in other religions.[20]

Two also contrasts with one, so that English proverb, 'second thoughts are best', has a Japanese equivalent. '*NIbamme no kangae ga saizen de aru*'. In contrast, two can also connote 'second best', and a poor second best at that. This is reflected in the irony of the Japanese, *NIkai kara megusuri wo sasu*, literally 'offering eye-drops from the second floor' – by implication to someone on the first floor – meaning that such help is worse than useless.

Three has any number of connotations in Japan. If most of these are favourable, one still notes the *SANsai* or 'three disasters' of fire, flood and storm. But then there are the *SANkei*, or three famous beauty spots, whose names at least, Matsushima, Miyajima and Ama-no-Hashidate, are known to almost all Japanese. Many of the expressions containing *three* associate things which only tradition brings into a single category. Thus the mirror, sword and jewel, which are presented to every new emperor on his accession are known as the *SANshu no jingi*, literally the three species of sacred articles.[21] At the profane level the *SANgyō* is a business 'composed of a restaurant, a house of ill fame, and geisha service' (Nelson 1974: 9).

Three can also be an extension of two, by adding a centre between two poles as in *SANze*,[22] which embraces the past, the present and the future. Another example of such extension is to be found in the word *SANiku*, which adds the hand to the heart-and-mind dichotomy implicit in *NIgedatsu*. This sense, in which the third element represents completion, is found also in the expression *SANdan rompō*, literally 'three-step argument', for a syllogism.

The examples just given suggest that the number 3, has great advantages as a basis for a taxonomy, and its use for this purpose is well established in Japanese Buddhism.[23] This explains why any number of entries in Mori (1980) are assigned to the category *bukkyō yōgo*, that is, 'Buddhist terminology'. Two examples are sufficient to illustrate this use. The first is *SANku*, the three forms of suffering: vicarious suffering,[24] suffering born of being deprived of pleasure,[25] and finally suffering caused by a lack of order.[26] The second example is much more fundamental; the principle expressed by *SANji*, literally 'three *times*', establishes the basic taxonomy. This succession of times – or *hōji* – begins with the entry of the Buddha into Nirvana, but different sects each have their own interpretation. Today, in popular usage, the *SANji* are equivalent to the *SANge*, and simply define the past, the present and the future, but in special contexts, such as agriculture, they have their own meanings, so that the first time, or season (*ji-setsu*), is that of sowing, the second, that of weeding, and the third, that of harvesting. This is significant because of the relation of these seasons, as I explain it in chapter 7, to the Buddhist part of the yearly calendar.

Not many Japanese proverbs incorporate the number 3 but no other. The two examples to be given point in different directions. The first, *SANnin, tora wo nasu*, means, roughly, 'three men can create a tiger'. The implication is that there is no tiger, but that if enough people say that there is one, they will come to believe in it. Any rumour, if repeated enough, will be believed. The second example, *SANzun no minaoshi*, means literally, 'look and correct three inches', so that at a certain level faults must be overlooked. There must always be room for some margin of error. *Sun*, the Japanese inch, is the basis of the word *sunpō*, meaning 'measure' or 'dimension', literally the 'law of inches'. The proverb is a gloss on this law, based on the principle that three *sun* is a small if not negligible measure. The contrast between the two proverbs lies then in the fact that the first implies that once three is reached, one is beginning to get into the realm of large numbers, whereas, according to the second, three still denotes a small number.

The process of looking at the *countable* numbers can be continued through the whole series, which is the procedure adopted by Mori (1980), but this would certainly distort the whole balance of this book, as well as trying the reader's patience. None the less, each number does have its distinctive attributes, and some of these are worth mentioning. The number 4, to start with, has any number of meanings which suggest symmetry, balance or completion. Some of these, such as *SHIji*, signifying the four seasons,[27] or by extension, the whole year round, have a meaning common to many other cultures.[28]

This corresponds, in the spatial domain, to *SHI-i*, meaning 'circumference' or *SHIkyō*, meaning 'boundaries', reflecting a traditional geocentric cosmology. In Japan, the land as so bounded, contains, in the social domain, the *SHImin*, literally 'four peoples' (or 'classes'), and by extension the whole nation.

In Japan there are still vestiges of a lore of four elements, or *SHIgenso*, which originated even earlier than that of the five elements, or *GOgyō*, examined in chapter 3. Now, almost all that is left is the principle that these four primordial elements, earth, water, fire and air, are at the root of all sickness. This is expressed in the phrase, *SHIHYAKUSHI-byō*, literally '404 ills'. The explanation is quite simple. Good health depends on the four elements being in harmony in the human body. If one gets out of harmony, it gives rise to a hundred ills, which become 101, if the discordant element itself is included. This, multiplied by 4, becomes 404, so that the '404 ills' comprise all human sickness (*Kanyō Kotowaza Jiten* 1988: 173).

Apart from its occurrence in *GOgyō*, the lore of the five elements, five governs any number of categories established in the Japanese culture, so that we have *GOtoku*, the five virtues, *GOsai*, the five colours, of *GOjō*, the five passions,[29] to give but three examples. The idea of there being five in such cases, may derive from *GOshi*, the five fingers of the hand.[30] In each case, the complete spectrum, within the category, is covered by its five components.

There are relatively few aphorisms containing the number 5. *GOho no shi* meaning, literally, a 'five step poem', means an astute answer, thought up on the spur of the moment. Yet more evocative is *GOri muchū*, literally 'in the middle of a five league fog', meaning to be completely at loss, unable to tell back from front. Here again one has the paradox that in the former aphorism, five essentially connotes a small number, while in the latter it connotes a large one. *Five steps* is nothing at all; *five leagues* is the whole countryside, but then of course there are a lot of steps in a league.[31] Once again context determines the meaning, illustrating once again the principle of metonymy.

Six, like four, can imply completeness, so that one has *ROPpō*, the six directions, constituted by adding up and down to the four cardinal points,[32] or *ROKUgō*, meaning the cosmos.[33] There is no obvious general principle governing the different connotations of seven. The *SHICHIyō* are the seven luminaries, comprising the sun, the moon and the five planets, Mars, Mercury, Jupiter, Venus and Saturn, known to the ancient world because their movement can be observed with the naked eye. These in turn provide the names of the seven days of the week (*yōbi*), as noted in chapter 3. One idiosyncratic case is the word *NANAtsuya*, for a pawnshop, which is derived as follows: *nana* and *shichi* are the alternative *kun* and *on* readings of the *kanji* signifying seven. *Shichi* is also the *on* reading of a quite different *kanji*, which with this reading means a 'pawn' or a 'pledge'. *NANA-tsuya* may be seen therefore as an example of the Japanese equivalent to Cockney rhyming slang. *SHICHIya*[34] is in fact an alternative word for a pawnshop, based on this *kanji*, but this word, in spite of its pronunciation is based on no more than a homophone for *shichi*, in its meaning of 'seven'.

The connotations of eight, an extremely auspicious number,[35] derive largely from its being the third power of 2. In the *ekikyō*, the art of divination based on the Chinese *I Ching* (which is considered in chapter 5), the eight *HAKke* are the basic signs. Like four and six, eight also signifies completion, to the point of perfection, as is reflected in the principle of *HAKkō ICHIu*, literally 'eight dimensions, one house', but taken to mean 'universal brotherhood'.[36] So also *HATtatsu* connotes 'all directions', so that *HATtatsu no hito* means a very broad-minded person. This also combines with *SHItsū*, or 'four roads', in the expression *SHItsū HATtatsu*, meaning 'accessible from all directions'. The arithmetical relationship to the number 4, here made explicit, is implicit in many other instances of 8, as this number is used in Japan. *YAegaki* means, therefore, 'fences within fences' on the implication that a normal fence encloses a square or quadrangle (or in Japanese *SHIkaku*).

In conjunction with *bu* or *bun*, in their meaning of a 'part', *hachi* can relate, implicitly, to *jū*, or 'ten'. The word *HACHIbume*, which can be used for its literal meaning, 'eight-tenths', can also be read as 'moderation'. In one expression, however, this reading is turned on its head: *mura HACHIbu*, literally 'village eight parts', refers to a household, with which social relations, expressed in terms of mutually rendered services, have been broken off. The needs of such a household will only be recognised in two extreme cases, fire and sickness, represented by the two parts still left out of ten after eight have been subtracted.[37]

Nine, in contrast to eight, seems to be rather an ominous number. There are the *KYŪsei*, the 'nine stars' used in predicting horoscopes,[38] the *KYŪsen*, the 'nine springs' of Hades, and the *KYŪten*, the 'nine directions' in heaven.[39] Then there is the saying, *KYŪshi ISshō*, literally, 'nine deaths one life',[40] which refers to a narrow escape from a dangerous situation. The implication is that there is only one chance in ten[41] of surviving. The implicit relation to ten occurs also in the expression *KUbu*, or 'nine parts', in the meaning of 'almost'.

Ten, not surprisingly, can also connote completion or perfection, but it does so in relation to the Chinese number system, adopted by Japan some 1,500 years ago as the basis for counting and arithmetic. To some degree it shares this attribute with the Chinese *man*, or 10,000, which is the culmination of the series of frame numbers, based on successive powers of 10 (and also adopted by the Japanese).[42] In both cases the use may be based on the meaning of a homophone. To take the case of *ten*, the Japanese *jū* is a component of the compound *JŪbun*, literally 'ten parts', whose normal meaning is 'enough' or 'sufficient', or even 'perfect'.[43] The same word can, however, be written with an alternative initial *kanji* (in its *on* reading), and meaning 'fill'. The *kun* reading provides the root for the verb *mi(chiru)*, 'to be full', or *mi(tasu)*, 'to fill', but these are normally written with yet another *kanji*, whose *on* reading is *man*. This, in various compounds, such as *mangetsu*, a 'full moon', also connotes completion. As already noted, this is also a property of *man*, meaning 10,000, in compounds such as *MANzen*, which means, simply, 'perfection'. This chain of transition, based successively on words with the same pronunciation, and words with the same meaning, is illustrated in figure 4.1. In numerical terms its result is that 10 = 10,000, which, if *prima facie* absurd, is still semantically acceptable.

'Ten' also occurs in the saying, *JŪnin TŌiro*, literally 'ten men, ten colours'; as such it is equivalent to the Latin, 'quot homines, tot sententiae'. But enough is enough, and the process of looking at the properties of low numbers must stop. Needless to say there are any number of special uses, in Japanese culture, of numbers between ten and 100, and some of these are considered elsewhere in this book.

*Figure 4.1* Why 10 = 10,000

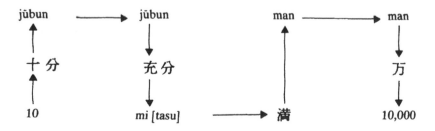

In particular, many such uses occur in the context of numerical institutions related to time, music and games. But now is the time to move
on.

## LARGE NUMBERS

In purely numerical terms it is impossible to identify any sort of
threshold separating *counting* from *large* numbers. The distinction
between the two is not a matter of arithmetic, but of cognition,
depending upon the circumstances of the case. The word 'centipede',
or its Japanese equivalent *mukade*,[44] both denote an insect with,
conceptually, 100 feet. Only a rather naive individual, whether
English or Japanese speaking, would actually believe that by counting
the feet one would end up at 100. The point is simply that one recog-
nises a centipede by the fact that its feet are too numerous to count.

Typically, 100 is the lowest of the large numbers, although ninety-
nine, in the expression *TSUZURA-ori*,[45] literally 'ninety-nine folds',
for a winding road, is also 'large' in this use. However, 100 occurs
much more frequently, in such common words as *HYAKka jiten* for
'encyclopedia' or *HYAKkaten*[46] for 'department store'. The occurrence of *sei* in HYAKUsei, meaning 'the common people', is more
allusive. The reference is to the 100 *common* names which ordinary
people share. It is derived from the Chinese *băixing*, the names of the
100 traditional clans to which all Chinese belong, and which strictly
have no Japanese equivalent.[47] Multiples of 100 are also used to
denote, simply, a large number, so *SAMbyaku* (300) can mean
'many', as can also *SANzen* (3,000). More specifically, *SAMbyaku
daigen* is an expression for a 'pettyfogging lawyer' (*daigen*), which
can be compared with *YAOya*[48] – the *800-shop* – meaning a 'green-
grocers', or by extension a 'Jack of all trades'. The contrast between
the negative connotation of 300 and the positive connotation of 800
in these two cases does not, apparently, have any further significance.

The connotation of abundance, or super-abundance, continues
with 1,000 (*sen* or *chi*), 10,000 (*man*) and still higher power of 10,
and their multiples. In the case of 1,000 there is the evocative
*CHIdori*, literally 'thousand bird', meaning the plover either because
of its movement, in a succession of small steps, on the ground, or
because of the sound of its wings (*haoto*) in flight (Mori 1980: 506–
7). The first usage leads to *CHIdori-ashi*, literally 'plover feet', to
describe the gait of a drunkard, staggering in all directions.

Normal usage of these higher powers of 10 is far less specific, but
there is almost always some difference of nuance, but with little syste-

matic basis. *SENnen*, or 1,000 years, means just that, but *MANnen*, 10,000 years, can mean 'eternity', so that, for instance, *MANnen yuki* means 'perpetual snow'. *Man* (or sometimes *ban*) is particularly auspicious, as in the familiar exhortation, *BANzai*, but there is always a counter-example such as *BANshi*, meaning 'certain death'. This latter case is significant in its implication of zero as an inversion of the connotation of infinity. *BANshi* implies a zero-chance of survival, but zero (*rei*) in the Japanese tradition is cognitively a much more difficult concept than 10,000,[49] or even *mugen*, or 'infinity'.

For the Japanese 10,000 is by no means the upper limit to large numbers. 千万, literally '1,000 10,000' (i.e. 10,000,000), in its *on* reading, *senman*, means just that, but the *kun* reading, *chi yorozu* means 'a very great many'. Drop a couple of million, so as to produce *yao yorozu* (literally 8,000,000), and this meaning is substantially unchanged. But one is still far from the limit. The numerical basis for *jūmanoku do*, an expression for 'paradise', is 10,000,000,000,000. This incorporates the next power of 10 beyond 10,000 (*man*) to have its own word, which is 100,000,000 (the square of 10,000), or simply *oku*. This is familiar to the Japanese from the phrase, *ICHI OKU nin*, literally 'one hundred million' ( 一億 ) people, which always means the whole population of Japan itself.[50] The significance of this usage is that it implies not infinity but a category which not only can be counted, but actually is counted by means of censuses – to which the Japanese are much addicted.

## CONTRASTING NUMBERS

The English-speaking world is familiar with expressions such as 'one in a million', 'at sixes and sevens' or 'two's company, three's a crowd', whose meaning depends on their containing two or sometimes more than two numbers. The popular use of numbers in Japan is dominated by such expressions. Some have the same meaning as familiar English sayings. An example is '*GOJUppo HYAPpo*', literally 'fifty steps, 100 steps', which is equivalent to 'six of one and a half dozen of the other'. The instances are, however, so numerous than many have no equivalent in the west, or at least none incorporating numbers. The problem is to find some sort of taxonomy for analysing them.

In numerical terms three different cases can be taken as the starting point. The first contrasts two numbers of a quite different order of magnitude – such as, indeed, 'one in a million'. The common phrase *man-ichi*, literally '10,000 ... 1', which is equivalent to 'if by any chance ...', provides a Japanese example. The proverb *IKka ni koto areba HYAKke isogawashi*, literally, 'if something happens in a house, 100 houses are busy', expresses the interaction between events summed up in the English, 'one thing leads to another'. The implicit social basis is typically Japanese. Much the same contrast between one and 100 occurs in the Japanese, *IKkenkatachi ni hoyureba HYAKkenkoe ni horu*, which is almost exactly equivalent to the English 'like dogs, when one barks all bark' with the typically Japanese use of a large number, 100, to connote 'all'.

One can equally be contrasted with a 1,000, or 10,000. An instance of the former case is *ICHIji SENkin*, literally, 'one [written] character [is worth] a thousand pounds[51] of gold'. The meaning is simply to affirm the unsurpassed value of the written word,[52] as first established by the Chinese classics. *ICHIji*, with a different meaning, represented by a different *kanji* for *ji*, is contrasted with *ban*, or 10,000, in *ICHIji ga BANji*, literally 'one thing 10,000 things'. This means that in the right context, one aspect of something is sufficient to disclose its entire character. An English equivalent would be, 'know one, and you know them all', particularly in a pejorative sense. The Japanese case is concerned more with the attributes of a single individual, as in the statement, 'you can see the sort of man he is from the way he ....', going on to add 'laughs', 'looks at you' or even 'parts his hair'. The implication then is that the man is not entirely to be trusted. This is a sort of proverbial application of the principle of *pars pro toto*.

The second case to consider, which is almost the exact opposite, is defined by the occurrence of two counting numbers. This further divides into three sub-categories. In the first of these the counting numbers are adjacent. The English example, 'at sixes and sevens,' has already been mentioned. In Japanese, any two adjacent numbers, in their *on* readings, can combine to mean either one or the other. This is commonly done to give a rough estimate of time, so that *ICHI-NIfun* is 'one or two minutes' and *SHI-GOnichi*, 'four or five days', but expressions such as *GO-ROKUnin*, 'five or six people' are also possible. This usage is less common for numbers greater than ten,[53] in which case only the units of the second number are expressed, so that, for instance, *JŪSHI-GOfun* is 'fourteen or fifteen minutes'.

Following the general case, there are a number of special cases. One of these is purely arithmetical: it is stated in the precept *SHIsha GOnyū*, literally 'reject four, admit five'. This is simply a rule for rounding off decimals to the nearest whole number, counting 0.5 and above as 1. The same paradigm is good for any number of non-arithmetical cases, so that in traditional medicine *GOzō ROPpu*, literally 'five viscera,'[54] six organs'[55] means all the internal organs. The two substantives can combine to form *zōfu*, which has much the same meaning. The numerical expression is significant for distinguishing five upper organs (*-zō*) from six lower organs (*-fu*), reflecting the opposition, upper–lower in its meaning pure–impure (Ohnuki-Tierney 1948: 37). The whole focus is on the abdomen, or *hara*, traditionally considered as the seat of the soul (ibid.: 58). At the same time, 5, an odd number is *yang*, while 6, an even number, is *yin*, which implies the same opposition.[56]

A similar distinction is to be found in *SHImoku SANsō*, or 'four trees, three grasses' (which is more common than the reverse order, *SANsō SHImoku*), although the binary opposition is less significant in this case. The expression refers to the seven specific plants whose cultivation was encouraged in every area of Japan during the Meiji era. The four trees, or bushes, are tea, mulberry, paper mulberry (*kōzo*) and lacquer (*urushi*), and the three grasses are flax, indigo and saffron (corresponding to the colours, red, blue and yellow).

In this sub-category, the two numbers can govern the same substantive, as in the saying, '*ROKUji SHICHIji no oshie*', which, in spite of the numbers occurring in it, is not equivalent to the English, 'at sixes and sevens'. If the literal meaning is 'teaching with six characters [or] seven characters', the actual sense is that a good way of life is best taught indirectly; the principle that example is better than precept is combined with that of following the path of least resistance – an essentially quietist, or better, Zen, approach to life.

In the next sub-category (which in Japan, at least, is probably more common than the second), the second number occurring is a multiple of the first. In one example, *GOJUppo HYAPpo*, already given on page 5, 100 occurs as a multiple of 50, and the occurrence of 2 as the multiplier is not uncommon. The classic case, represented by the English 'killing two birds with one stone', is literally translated, in inverted form, by the Japanese 'ISseki NIchō', whose proverbial sense is the same as that of the English version. In Japan another popular saying is 'GOfū JŪu', literally 'five wind ten rain', and by implication, wind every five days, and rain every ten. This, by Japanese standards, is the ideal climate, so the expression refers to a state of affairs in which everything goes right. The metaphorical expression, 'sailing with the wind', is an English equivalent.

Many numerical expressions derive from the different traditions of Japanese Buddhism. In the present category the 5–10 opposition occurs in 'GOkyō[57] JISshū', meaning 'five faiths ten sects'. This is a precept of Kegon, the major sect of Mahayana Buddhism in Japan, which has been prominent ever since it was established in the eighth century. The implicit meaning is reductionist. Ten can be reduced to five, and by implication, five to one. This accords with Kegon doctrine, which conceives of the whole world being present in every speck of dust, so that all things pervade each other. This is the principle also of many of the expressions, already considered, in which one is contrasted with a very large number, such as 10,000.

The miscellaneous sub-category in which the two numbers occurring are not related to *each other* arithmetically is inevitably heterogeneous. Sometimes the combination imports an arithmetical operation. Two examples using multiplication, and one using addition, illustrate this point. The first is *minanoka*, literally '3-7 day', and referring to the twenty-first day falling after a death, as well as the Buddhist rites then to be performed.[58] The importance of the number 7, is that it shows that these rites belong to a series, all occurring at varying intervals of seven days, of which *shonanoka*,[59] performed after seven days, and *chūin*, after forty-nine days, are the most important. The second example is *SHIROKUjichū*, or '4-6 hour duration', that is 'throughout the twenty-four hours of the day', and so, by derivation, 'constantly' or 'always'. It is no coincidence that these two examples are both concerned with the lapse of time (for reasons to be given in chapter 7), but this is not true of the third example, based on addition.

*SANbu wa takumi, SHICHIbu wa shujin*, literally translated into English, is 'three parts craftsman, seven parts master'. Now $3 + 7 = 10$, and ten parts, *JŪbun*, is, as already noted, completion or perfection. The contribution of the master, who makes the design, scores 70 per cent as against the 30 per cent scored by the craftsman who carries it out. This is a statement about the hierarchy of craftsmanship.

Measurement in the same units is also to be found (as is also esteem for craftsmanship) in *GOryō de obikōte SANryō de kukeru*, or 'an obi bought for 5 ryō is blindstitched for 3 ryō'.[60] The *obi* is simply the sash which many Japanese, particularly women, wear around their waist. At first sight it is nothing more than a long piece of cotton cloth, so that the implication is that 5 ryō is rather a lot to pay for it. The hidden craftsmanship, represented by the blind-stitching, which is invisible from the side of the sash worn outside, is, however, worth 3 ryō.

A final example also combines three with five, used, however, to count different units of measurement, which is essential for the meaning of the proverb. *SANzun no shita ni GOshaku no mi wo korobasu* means 'three inches of tongue destroy five feet of body'. The meaning is that a man can be destroyed by a little bit of idle talk. Paradigmatically this is close to the English, 'an ounce of prevention is worth a pound of cure',[61] but the sense is closer to 'do not spoil the ship for a ha'porth of tar'.[62]

## THREE NUMBERS AND MORE

Who has not heard of the festival of *SHICHI-GO-SAN*, or 7–5–3, at which girls of 3 and 7, and boys of 5, dressed in their best new clothes, are presented at the local Shinto shrine? And is it coincidence that the sum of these three numbers is 15, and the ceremony takes place on November 15? This is almost certainly the best-known three-number combination in Japan, and there do not seem to be so many others. Shinto also provides *SAN–SAN–KUdo*, or '3–3–9 times', referring to the ceremonial exchange of cups of *sake* at a wedding, which takes place in three phases, each with three ritual exchanges. *Sake* occurs explicitly in the proverb, *IPpai wa hito sake wo nomu, NIhai wa sake sake wo nomu, SANpai wa sake hito wo nomu*, or, literally, 'one cup, the man drinks the *sake*, two cups, the *sake* drinks *sake*, three cups, *sake* drinks the man'. This is no more than an elementary plea for temperance.

Finally, there is a four-number combination which provides a rule for interpreting as a portent the number of times one sneezes. '*ICHI soshiri NI warai SAN hore SHI kaze*' does not actually mention sneezing. The message is that one (sneeze) means being insulted, two, being laughed at, three, being admired, and four that one is simply catching a cold. This is a good note for moving on to the final section of this chapter.

## MEANING AND FORM IN KANJI NUMERALS

This final section interprets the written form of the numbers 10, 77, 88 and 99, all of which are regarded by the Japanese as special cases. The *kanji* for 10, 十, is to any westerner simply a cross, but to a Japanese a cross is the number, 10, or 十. The *kanji* is then simply *JUji*, or 'ten character', and in concrete form this becomes *JUjika*, or 十字架, the final character representing a *frame*.[63] This is then the word Japanese Christians use for the cross of Christ. The written forms of 77 and 88, can be reconstructed to represent the first *kanji* in *kiju*, 喜寿, and *beiju*, 米寿, meaning, respectively, a 77th and an 88th birthday. The *kanji*, 寿 for *ju*, in its *kun* reading, *kotobuki*, means 'a long life' or 'congratulations for a long life'. All the combinations in which *ju* occurs are highly auspicious. The *kanji* 喜, in *kiju*, combines the component *kanji* in 七, 十, 七 for 77, to form a single character, 喜, for *ki*, meaning to 'rejoice'. If this process is not very transparent, there is much less difficulty in seeing how the *kanji* 八, 十, 八 for 88, combine to form 米, or *bei*, for 'rice' (which in Japan is highly auspicious). Finally *hakuju*, for a ninety-ninth birthday, arises as a result of a quasi-arithmetical process. If from 百, for 100, 一, for 1 is subtracted, the result is 白, which in its *on* reading is *haku*, and in any case means 'white'. Arithmetically, $100 - 1 = 99$, so that *hakuju* is taken to represent a ninety-ninth birthday.[64]

All these representations are common ground to the Japanese. Whether the same is true in the instances looked at in the preceding sections depends on the particular case. The principles, in any case, are familiar to all Japanese, and the instances given are but a small fraction of those recorded. The principle is well established that in a highly numerate culture, such as that of Japan, a high level of diversity, and of creativity as well, must be expected in the use of numbers.

# 5 What's in a Japanese name?

## NAMES AND NUMBERS

Before looking at the part played by numbers in Japanese names, it is essential to realise that the fact that such names are almost invariably written in *kanji*[1] automatically attributes to them the meaning of their component characters.[2] There are, however, so many deviant readings of *kanji* occurring in proper names, that special dictionaries are needed to list them.[3] This means that any Japanese consulting a map, timetable or telephone directory will be at loss as to how many of the names occurring should be pronounced, even when the characters comprising them are themselves elementary. One reason for this is that names, or proper nouns often have spoken forms which, if not obsolete, are confined to this form of speech. If few people in England stop to think that Sutton means 'south town', its meaning, in any possible spoken variant, would be made clear to a Japanese by the *kanji* comprising its written form,[4] and this is a comparatively simple case. True, in such transparent cases as Redhill, an Englishman will not think of a *red hill*, and nor will a Japanese, hearing the name Hiroshima, think of a *wide island*. None the less, the written form of Hiroshima, 広島 , will remind a Japanese of this meaning in a way without any equivalent in western languages. In relation to numerals, the point is extremely important, since any advanced culture tends to conceive of numerals as being pre-eminently within the domain of visible language.[5]

   In contrast to the English-speaking world, where the use of numbers in names is uncommon,[6] proper names in Japan, both of people and places, are suffused with numbers. It is difficult to see any rhyme or reason in the countless instances which occur, beyond the fact that such number-names are often descriptive. Examples are Kyushu and Shikoku, two of the four main islands of Japan. Both *shū* and *koku* represent the states or districts into which the two islands were divided, there being nine (*kyū*) such units in Kyushu, and four in Shikoku.[7] The *shū* of Kyushu occurs also in Honshu, the main island of Japan, but this name connotes no more than 'main land', or perhaps the source of all *shū* – which is now, in contrast to *kuni* or *koku*, a somewhat archaic word. Hokkaido, the fourth of the main islands,[8] also has a number-based name according to the scheme of chapter 3. The three components of the name, *hoku, kai, dō*, mean 'north sea province', so that *hoku*, 'north', is one of the four numbers in the cyclical system of the cardinal points. At the same time, in the scheme of the five elements, the centre was treated as belonging to the cardinal points,[9] and this is the essential meaning of *hon* in Honshu. This cannot be taken to mean that Honshu and Hokkaido represent two vestigial points in some primordial geography of Japan, in which, originally all the five elements were represented. The truth must be that Japanese place names represent the surviving elements of many different models, current at an earlier stage in Japanese history. Today the part of the country lying on the Japan Sea to the north of the Noto peninsular is known as *hokuriku*, or simply 'north country', but there is no corresponding *nanriku*, or 'south country'.[10] Along the east–west axis there are the more familiar Kan-tō and Kan-sai, the former containing the eastern capital, Tō-kyōto,[11] and the latter the western capital, Sai-kyōto. The latter, the actual capital of the country until the Meiji restoration in 1868, is still known by its original name, Kyoto,[12] for the name, Saikyo, then proposed to distinguish the city from the new capital, Tokyo, never caught on.

Place names defined by political boundaries are by no means confined to the cities of (Sai)Kyoto and Tokyo(to), but other instances tend to occur much more haphazardly. An example is Mikuni, just outside Osaka, in which the written form, 三国, immediately recalls that of the island of Shikoku, 四国. This is a case of a double transformation: Mikuni,[13] the point common to three of the ancient *kuni*,[14] represents not only a change from a topology based on areas to one based on points, or vertices, but also a lexical change from an *on* to a *kun* reading.

A mere catalogue of all the hundreds if not thousands of Japanese place names incorporating numbers would, on the face of it, create nothing but confusion. A limited number of further examples is, however, sufficient to illustrate the semantic possibilities, as well as the problems which arise in interpreting them. As for the meaning of place names, it is useful to start with six quite familiar instances taken from Tokyo,[15] Futako, Mitaka, Gotanda, Roppongi, Kudanshita and Chiyoda, incorporating, respectively, the numbers 2, 3, 5, 6, 9 and 1,000.

On the basis of this numerical order, and so beginning with Futako, this proves to be a suburb in the western part of the city. The name has two possible meanings, both derived from the combination of *futa* ('two' in its *kun* reading) and *ko* (child). The name could simply be that given to a girl, in principle a second daughter.[16] Alternatively, it could simply mean '*two* children', disregarding, in this case, the ordinary rules of syntax.[17] In either case the name would have been adopted for some forgotten historical reason for a place later to be swallowed up in twentieth-century Tokyo.

Mitaka, known to most Tokyo residents as a station of the Chūō Main Line, means no more than 'three hawks'. It is difficult to say anything about this name, save that its origin is almost certainly to be found in some local legend, dating from the time when the sport of falconry was still a part of country life in Japan. In this sense it is less enigmatic than Futako, although it may be equally old.

Gotanda is a name which would be evocative in any agricultural community. Its meaning, roughly translated, is 'five-acre field', which is just the sort of name one might find given to a field in an English farming village. It matters not that a *tan* measures but a quarter of an acre, nor that *da* (or *ta*) means, quite specifically, a rice or paddy field.[18] If Gotanda is now familiar to most residents of Tokyo as one of the stations on the Yamanote railway (which is roughly equivalent to London's Circle Line), there is no problem about how the name first arose. The same, or similar, names must occur throughout Japan.

Roppongi (a high-class residential area favoured by foreign businessmen) is equally evocative, but in rather a different way. Its lexical base is a little confusing,[19] but if it has a meaning, it is simply 'six trees', recalling the English Sevenoaks.

Kudanshita, literally 'nine steps down', is a part of Tokyo close to Sanbancho and the Yasukuni shrine. The inclusion of the word *dan* makes the name significant. If the primary meaning of *dan* is 'step', the common connotation is of ranking in a hierarchy, determined on the basis of achievement and competition, such as that which determines the status of sumo wrestlers or players of the game go. In the case of Kudanshita the primary meaning of 'step' is almost certainly correct, for reasons which I give in chapter 8. The question remains, why *ku*, or *nine* steps? Names with other numbers of steps do not seem to occur,[20] and I have not been able to do better than guess at answers.[21]

Finally, there is Chiyoda, the *ku,* or city ward, in which the imperial precinct is situated. The name, literally 'thousand ages rice-field',[22] could hardly be more auspicious. Mori (1980: 507) gives the meaning of *chiyo* as an 'exceptionally long period of time', and refers to a poem with this title, in which a man's prayer for a long life is answered, and he is congratulated upon his achieving it. A rice-field, maintained over a period far beyond any human memory, represents also an achievement which only constant devotion could have assured. Chiyoda, to any Japanese who considers its meaning, represents an ideal of continuity and stability in the most important of life's activities, the cultivation of rice. It seems hardly a coincidence that the emperor, whose well-being is closely associated with the celebration of the annual harvest of rice, lives at the heart of Chiyoda.[23]

*Chi,* with its favourable connotations, probably occurs in more place names than any other number. It occurs, for example, in Chiba (literally, 'thousand leaves'), one of the only two names of a prefecture incorporating a number.[24] But of all the names in which it occurs, Chiyoda is probably the most auspicious.

Having started with Tokyo, it would be agreeable to travel throughout the whole of Japan, noting yet more curious examples of place names containing numbers. The resulting catalogue would be, however, far too long. But it is worthwhile to note two towns on the coastal railway along the inland sea just to the west of Hiroshima. The first of these is the 'five-day market', Itsukaichi, and the second, the 'twenty-day market', Hatsukaichi. Both these names incorporate the distinctive forms for counting days given in chapter 2. The significance of the names is entirely transparent and their origins not difficult to discover in the local history.

The journey through Japanese place names can appropriately be brought to its end by looking at two instances which are anything but transparent: one is the Osaka district of Juso, and the other, the Tsukumo bay on the Noto peninsular on the Japan Sea coast.[25] Of these two names, the former means simply 'thirteen' and the latter, 'ninety-nine', but both are in non-standard forms, and any Japanese from outside the area concerned would be at a loss for the correct pronunciation. Juso appears to be the only instance of 'thirteen' occurring in a place name. 'Ninety-nine' however, also occurs, in its standard form, in the name, Kujūkuri (O'Neill 1972: 4).

It is now time to turn to the names given to individuals. This requires a distinction to be made between given and family names. The latter, before the modern era, were almost unknown among ordinary Japanese. In 1872, however, the formal distinction between *daimyō,*[26] samurai, peasants and merchants was abolished, and the use of family names became universal. Although some are highly idiosyncratic, the majority have a simple form, written in two *kanji,* generally in the *kun* reading.[27] The most common semantic base is topographical, so that *kawa* (river), *mura* (village), *ta* (rice-field) and *yama* (mountain) constantly occur, in such familiar names as Kawaguchi, Kimura, Tanaka and Yamamoto. These commonly combine with the words for the cardinal points, giving, for example, Kitamura (north village). Equally common are the five elements, fire, water, wood, metal and earth, so that one finds Shimizu (from *mizu,* 'water') and Kimura (from *ki,* 'wood'). Finally the number-words themselves are often part of a family name. All three categories can combine in a single name, such as Kindaichi, meaning 'metal rice-field one'. As with place names, deviant readings also occur, so that one encounters 五十子, or Ikago, meaning 'fifty children' (for which a standard reading would be *gojūko*).[28]

The importance of a name in determining the character and destiny of an individual has long been accepted in Japan. Although I focus my discussion on the choice of names at birth and, in the case of women, at marriage, there are other occasions when a name can be changed, generally to take into account an important change in circumstance. Fukuzawa (1966) gives any number of instances of samurai known to him, who, after the Meiji Restoration of 1868, chose new names to accord with their new position in life. *Sumō* wrestlers, once they have been accepted into the ranking system which I describe chapter 10, almost always acquire new names to give them the right public image. And almost all Japanese are given a new name after their death.

Where the majority of Japanese family names fall within a relatively narrow range of possibilities, the choice of a given name is almost unrestricted, particularly for boys. True, there are popular boys' names, such as Takeshi, and many of these are appropriate to a son according to the order of his birth, so that a second son, for example may well be named Jiro, of which the written form, 二郎, makes the meaning clear, but parents are equally free to contrive a name which they believe to be auspicious. Girls' names tend to end with, *ko,* simply meaning 'child', but this also leaves considerable scope at least for choosing the first part of the name. Number-words, such as in Nanako (七子),[29] probably occur as frequently as they do in boys' names, in which case the problem of choosing the name is simply solved.

Although it is permissible to choose a name according to the sort of aesthetic criteria which often govern such a choice in the English-speaking world, the Japanese are much more inclined to choose a name for its own particular meaning. The popular Takeshi is a case in point. The first component *kanji*, 武, connotes precisely the virtues[30] which Japanese parents would like a son to have as he grows up. But even if such a choice is made, it may still, before it becomes definitive, be subjected to numerical tests which are so involved that they are the subject matter of a special branch of learning, whose interpretation and application demands considerable expertise.

## THE FULL-NAME SCIENCE OF SEIMEIGAKU

The numerical dimension in choosing appropriate names goes far further than just looking at the meanings of the component elements as given in a dictionary.[31] In practice the choice of such names, which may be entirely idiosyncratic, will not be determined by what is to be found in the dictionary. It is then *seimeigaku*, or 'full-name science', which guides the choice. This esoteric art can only be applied with the help of manuals, of which a wide selection is available in almost any supermarket or station bookstall. Parents who find the manuals too difficult (or who cannot agree with each other about how to apply the rules) can resolve their difficulties by consulting any one of the numerous *ekisha* listed in the yellow pages (and whose work I describe in chapter 6).

What, then, are these rules? The answer to this question can best be stated as follows: the rules establish, first, a number-based taxonomy, according to which the name of any Japanese individual can be analysed, and second, a procedure for combining the numerical results of this analysis to yield a set of numbers, which, under different heads, reveal the character traits to be expected from anyone to whom the name has been given. The procedure is but one more instance of the common practice of relating a number to destiny.[32] In this sense *seimeigaku* is an oracle, which may be consulted, and whose advice, often ambiguous, then falls to be interpreted and acted upon. How this works in practice can best be illustrated in an example such as is given in figure 5.1.

The starting point is the *kanji* version of the name. The example chosen relates to a girl, so that the analytical process is carried out twice, with the results numbered 1 and 2. Case 1 relates to the choice of the name Shigeko, given to a girl born into the Tanaka family. Case 2 relates to a later stage in the life cycle, when a husband must be found for Tanaka Shigeko.[33] This case contrasts with the first, in

*Figure 5.1* Seimeigaku

| | | A | B | C | D |
|---|---|---|---|---|---|
| 1. Name in *rōmaji* (unmarried = 1, married = 2) | 1 | Ta – | Naka | Shige – | ko |
| | 2 | Take – | Mura | | |
| 2. *Kanji* equivalent of 1 | 1 | 田 | 中 | 子 | 子 |
| | 2 | 竹 | 村 | | |
| 3. Number frame | | A | B | C | D |
| 4. Number of strokes | 1 | 5 | 4 | 3 | 3 |
| | 2 | 6 | 7 | | |
| 5. *Yin-yang* of 4 | 1 | yang | yin | yang | yang |
| | 2 | yin | yang | | |
| 6. Binary equivalent of 5 | 1 | 1 | 0 | 1 | 1 |
| | 2 | 0 | 1 | | |
| 7. *On* reading in *rōmaji* | 1 | *den* | *chū* | *shi* | *shi* |
| | 2 | *chiku* | *sou* | | |
| 8. *Katakana* initials of 7 | 1 | テ | チ | シ | シ |
| | 2 | チ | ツ | | |
| 9. '8' based on five elements | 1 | fire | fire | metal | metal |
| | 2 | fire | metal | | |
| 10. Quinary equivalent of 9 | 1 | 3 | 3 | 2 | 2 |
| | 2 | 3 | 2 | | |

| Ten signs under five heads | | Status | Total | Sign |
|---|---|---|---|---|
| First head (*tenchi* A > or < B) | | 1 | | + |
| | | 2 | | – |
| Second head: literal meaning of name | | 1 & 2 | | + |
| Third head: totals from stroke count | | | | |
| A + B (*ten'i*) inherited traits | | 1 | 9 | –/+ |
| | | 2 | 13 | ++ |
| B + C (*jin'i*) acquired traits | | 1 | 7 | + |
| | | 2 | 10 | – |
| C + D (*shonen*) first years of life | | 1 | 6 | ++ |
| | | 2 | 6 | ++ |
| A + B + C + D (*chi'i*) last years of life | | 1 | 15 | + |
| | | 2 | 19 | – |
| A + D (*fukuon*) qualifies first four | | 1 | 8 | + |
| | totals | 2 | 9 | – |
| Fourth head: *yin-yang* | | 1 | | + |
| | | 2 | | – |
| Fifth head: (*gogyō* or five elements) (rank) two middle numbers | | 1 | | + |
| | | 2 | | – |
| (destiny) all numbers | | 1 | | – |
| | | 2 | | – |

that the given name, Shigeko, cannot be changed, while there is no obligation to accept the family name of a prospective husband, if the *seimeigaku* oracle is unfavourable. All that is then needed is to continue the search.

In both cases the analysis begins with the *kanji* version of the name. The first step is simply to count the number of strokes, which, being standard for every name, provides the basis for its listing in any Japanese dictionary (as already explained on page 31). In the elementary case (such as the example given in the text), the stroke-count yields two numbers for each part of the name, but this is not essential. For the many Japanese names which cannot be broken down in this way, there are prescribed means for modifying the procedure illustrated.

The four numbers, represented algebraically by $A$, $B$, $C$ and $D$, each have (5) a *yang* or *yin* sign according to whether they are odd or even, and the binary equivalent (6) is then 1 or 0.

The next stage is to find the equivalent or $A$, $B$, $C$ and $D$ in the quinary system, known as *gogyō*,[34] of the five elements. This is a somewhat involved operation. The component *kanji*, $A$, $B$, $C$ and $D$, will each have an *on* reading[35] (7), but in most cases it will not be correct for the actual name. None the less, the initial *kana* of each of these readings (8), will, according to the scheme given in figure 5.2, correspond to one of the five elements (9). Each element, in turn, has its own number from 1 to 5 (10).

The actual interpretation is based on the concepts of *yoshi* (+) and *kyō* (−), the former connoting everything that is auspicious, and the latter, everything that is inauspicious.[36] The first use of this dichotomy, under the first head, *tenchi*, is elementary. It determines, according to whether $A$ is greater or less than $B$, the status of the name according to the dichotomy, heaven (*ten*) and earth (*chi*). The second head has no numerical basis:[37] it simply assesses the name according to its literal meaning, which will often be what first led to its choice – subject to further examination according to the canons of *seimeigaku.*

The third head, based on different totals derived from the stroke-count given in line 4 of figure 5.1, is much more involved. In this way the example given produces (twice over) five totals, $A+B$, $B+C$, $C+D$, $A+B+C+D$ and $A+D$:

1.    9,  7,  6, 15 and 8, and
2.    13, 10,  6, 19, and 9

Row 1 provides the basis for interpreting the name actually given at birth to Tanaka Shigeko, while row 2 does the same for the name she

*Figure 5.2* Gogyō equivalents of the kana syllabary

| 木 Wood | 火 Fire | | | 金 Metal | 水 Water | | 土 Earth | | |
|---|---|---|---|---|---|---|---|---|---|
| カ ka | ラ ra | ナ na | タ ta | サ sa | ハ ha | マ ma | ア a | ヤ ya | ワ wa |
| キ ki | リ ri | ニ ni | チ chi | シ shi | ヒ hi | ミ mi | イ i | イ (y)i | ヰ (w)i |
| ク ku | ル ru | ヌ nu | ツ tsu | ス su | フ fu | ム mu | ウ u | ユ yu | ウ (w)u |
| ケ ke | レ re | ネ ne | テ te | セ se | ヘ he | メ me | エ e | エ (y)e | ヱ (w)e |
| コ ko | ロ ro | ノ no | ト to | ソ so | ホ ho | モ mo | オ o | ヨ yo | ヲ (w)o |

would acquire if she were to marry into the Kitamura family. The actual meaning of every number is found by looking in the list given in the manual, which places every number from 1 to 81 in one of four categories,[38] based upon *yoshi* and *kyō*. The four categories are then *daikichi*,[39] meaning 'excellent', *yoshi*, meaning 'lucky', *hankichi-hankyō*, meaning 'half and half', and simply *kyō*. The whole scheme has a clear rationale. The inherited traits (*ten'i*) are determined entirely by the family name, simply comprising $A$ and $B$; the acquired traits (*jin'i*) take one element from the family name, $B$, and one, $C$, from the given name; the first years of life (*shōnen*), during which a girl, in particular, may actually be addressed by her given name, are determined by the two components, $C$ and $D$, of this name, while the last years of life (*fukuon*) are determined by the total, $A+B+C+D$. As logic demands, the first years of life, represented by the third number in each row, is the same in both. Plainly a girl's fortune in the first years of her life cannot be influenced by the choice, not yet made, of her future husband.

The interpretation of the fourth head, *yin–yang*, and the fifth head, *gogyō*, depends upon an analysis of the order of the components, which in the former case take the form of a binary, and in the latter, of a quinary number. The same rule applies in both cases. The basic principle is that all combinations are *yoshi*, except for four[40] definite cases, in which they are *kyō*:

(i)  *Ichiritsu* (in the same way). All the signs are the same. This is obviously much more likely to occur with the two elements of *yin–yang* than with the five elements of *gogyō*;

(ii) *katayori* (one-sided). All the signs are the same, except for the first or last. This explains why both *yin–yang* in row 2 and destiny in row 2 are *kyō*;

(iii) *wakare* (separation). The first two signs are the same, as are the last two. This explains why destiny in row 1 is *kyō*.

(iv) *hasami* (scissors). The outermost signs are the same, as are those contained between them.

The only possible conclusion is that Tanaka Shigeko, by marrying into the Kitamura family, will only change good fortune for bad. Given her family name, Tanaka, the name, Shigeko, was well chosen for her by her parents. If, then, they were to continue to follow the path of wisdom, they should eschew a Kitamura son-in-law.

This is, however, only one side of the whole matter. There is a separate department of *seimeigaku* dealing with *aishō*, or compatibility between marriage partners. Although Tanaka Shigeko would seem to be ill-fated in the Kitamura family, it could still be that the marriage itself was auspicious according to the canons governing *aishō*. These are relatively simple. The starting point is the first *kanji* in the given name of the two partners (it being assumed that after the marriage both will have the same family name, that of the man).

First, on the principle of *yin–yang* (based on an even or odd number of strokes), the two signs must be different. Giving the prospective Kitamura bridegroom the name, Takeshi, this requirement is met, for the *kanji* for *take* has eight strokes in contrast to the three strokes for *shige*.

Second, the requirements of *gogyō* must be met. These are rather more complicated. The *on* reading equivalent to *take* is *bu*,[41] which falls under the element water, where *Shige* (from figure 5.2 on p. 71) falls under metal. The question then is, are water and metal compatible? The answer is negative. This combination falls under the rubric *tagai ni koroshiau mono*, literally, 'people who kill each other'.

In most cases this would be the end of the matter, or so one would think. But not every family pays equal attention to the oracle. Circumstances can alter cases. For Tanaka Shigeko's father, marrying his daughter into the Kitamura family could be good for his promotion prospects at work. Things could still work out for the best.

The lonely and unhappy housewife is, however, a well-known figure in modern Japan, and her misfortunes have not been overlooked by the leaders of the new religious sects which have sprung up. Risshō Kōseikai, one of the many sects[42] adhering to Nichiren Buddhism,[43] which emphasises the importance of ancestor worship, has a particular appeal to married women who are allowed to choose a new name the better to achieve this end. Risshō Kōseikai still distributes its own manual (Kobayashi 1951) to new converts,[44] although this part of its appeal is now somewhat played down. In the example given above, Kitamura Shigeko, as a married woman, would be free to have a new name given to her in place of Shigeko,[45] making sure that any possible choice be first tested against the canons of *seimeigaku*. The changes then arising in the numbers, *C* and *D*, would in turn change *shōnen*, governing the first years of life, but leave *ten'i*, representing the husband's inherited traits, unaltered.

## THE LOGIC OF JAPANESE NAMES

The familiar Japanese visiting card, or *meishi*, is the most common representation of an individual's full name, or *seimei*. An individual, in the course of a life in which exchanging such cards is an everyday event, will assemble a collection expressing the whole gamut of his social and business relations. Towards the end of the year, a decision must be taken as to which of the individuals represented by the cards will be sent a new year's card, thus defining for the coming year the recognised circle of acquaintances. The names will be as stated on the *meishi*, but will otherwise be no part of the greeting. Japanese do not generally use any form equivalent to 'Dear Mr Brown' or 'Dear Jack'; a card or a letter will start with a perfectly formal greeting, such as *haikei*, which does little more than express respect for the person addressed.

Ordinary conversation does allow the formal 'Tanaka *san*', or for a teacher, 'Tanaka *sensei*', but the honorific terms are never used in introducing oneself. A telephone caller, after establishing contact with the formula *moshi moshi*, may simply announce his name with *Tanaka desu*, literally 'It is Tanaka'. The use of given names in such a context is not at all usual; the most common case is perhaps that of a daughter calling her mother. In ordinary conversation such use is confined to the family, and then only to cases when a senior addresses a junior member (and an older brother or sister is definitely senior to a younger).[46]

The point is that the world of *seimei* is a 'third person' world, in which the full name is a sort of surrogate for the individual. It is as if Kitamura Takeshi, in presenting his *meishi*, is announcing that he is an actor playing the part of Kitamura Takeshi, with all necessary details of the role stated on the card. As an individual he in turn confronts a stage on which the *dramatis personae* are defined, in turn, by his collection of *meishi*. It is a world, however, in which any reference to himself, allowed by the language he speaks, is ambiguous.[47] The loss of an explicit demonstrative, referring to himself, such as the unequivocal English, 'I',[48] leaves the individual Japanese disembodied, so that he must take a detached view of himself, which is precisely what is represented by his *meishi*.

In mathematical terms, the *meishi* reduces the individual to a coordinate, to a point on a graph, and in my view this provides the explanation, not only for the readiness of Japanese to include numbers in names, but also their obsession with numerical oracles, such as *seimeigaku*. This is borne out by the form of Japanese ancestor worship in which the destiny of any Japanese is to be represented by a tablet (*ihai*) on the domestic altar (*butsudan*), which becomes a sort of post-mortem surrogate for the *meishi*.[49] The domestic altar belongs to the household, or *ie*, which in the modern era is identified by the family name. If then, as I show in chapter 7, the temporal cycle of this life continues into the next, the choice of the correct name for this life is, in terms of individual destiny, as critical as is the choice of a name for the life after death. Whereas, however, the choice of the latter name is left to the priests of the Buddhist temple which looks after the mortuary ritual,[50] the choice of the former name is a matter to be decided within the family.

The choice of a name in the twentieth century is a compromise with a bureaucracy which could not cope with any uncertainty about an individual's identity.[51] This is a recent development. Chamberlain (1974), writing at the beginning of this century, pointed out how

> it was formerly the custom of a man to alter his name at any crisis of his career. Even now adoption, and various other causes, frequently entail such changes .... Not human beings only, but places exhibit this fickleness .... The idea, which is an old Chinese one, is to emphasise by the adoption of a new name some new departure in the fortunes of the city, village, mountain, school, etc. (Chamberlain 1974: 347)

No wonder, then, that the chance of adopting a new name still remains a part of the package offered by Risshō Kōseikai. The art of *seimeigaku*, in its present form, is not so much a survival from a superstitious past, but an adaptation of an ancient institution to the exigencies of the modern age.

# 6 Fortune-telling

## DIVINATION

According to Yoneyama (1990: 172) divination is the means of answering questions on matters requiring a decision about future policy or action, when there is no sufficient rational or logical basis for finding an answer. The meaning of the word he uses for divination, *uranai*, is entirely general and can apply to the interpretation of dreams, the entrails of sacrificed animals, the movements of birds, and all kinds of phenomena, whether natural or contrived. The questions to be answered are just as varied. They can relate to the finding of lost objects, the settlement of disputes, the treatment of sickness, the planning of a journey or a military campaign, the choice of a daughter-in-law, to mention some of the examples given. The world in which *uranai* was established was that of traditional rural Japan, where concern for the harvest or success in fishing or hunting were dominant issues in everyday life.

Although the means of divination were not necessarily numerical, it is significant that one word for a diviner, *uraya-san*, in its second component, *san*, imports counting or calculation.[1] This is certainly true of the form of divination known as *eki*, which refers particularly to the *I Ching*, or *Chinese Book of Changes*, known in Japan as the *ekikyō*. However it may have been in the past, there is no doubt that *uranai*, as at present practised, is often based on numbers. This may be the result of two related factors in the recent history of Japan: migration from the countryside to the town and mass literacy (Yoneyama 1990: 172). In this chapter I consider, apart from *ekikyō*, two other types of popular divination, *omikuji* and *kyūsei*. The latter, which literally means 'nine stars', is the normal word for astrology or horoscopes. The former word, *omikuji*, is untranslatable, but its meaning will become clear enough in the following section.

# OMIKUJI

*Omikuji* is a popular institution,[2] to be found in almost every shrine and temple in Japan, although, as with many such institutions, its origins are certainly Chinese (Ronan and Needham 1978:194). In the sense of material culture it has two components. The first is a metal box, something less than a foot in length, with a hexagonal cross-section, of which each side measures about an inch. At one end there is a small hole. The contents consist of a number of metal rods, not quite so long as the box itself. If the box is held upright, and then shaken, there is sufficient space for *one* of these rods to come through the hole. A metal boss at the other end, such as one finds on a knitting needle, prevents it falling out. Each of the rods, which may be regarded as the essential *kuji*, bears a different number. This is the basis of the *sacred* lottery, used to determine the will of the gods.

The other material component consists of an upright frame, about the height of a fully-grown adult, supporting a matrix of pigeon-holes, each of which is designated by a combination of two numbers; of these, one, corresponding to the row of the matrix, is that on the *kuji*, while the other is based on the date on which the oracle is consulted. Each pigeon-hole contains a supply of leaflets, in size rather smaller than a post card, on which the message of the gods is printed. The leaflets in any one pigeon-hole will be identical, but different from those in all the other pigeon-holes.

Nothing could be simpler than consulting the *omikuji* at one's local shrine or temple, so it is not surprising that the institution is extremely popular with children. For a small donation in a collecting box, one is free to shake the box and obtain a number. This process, known as *kujihiki*, or 'drawing a kuji', allows one to take the corresponding message from the rack. If this is favourable, all well and good, if not, then the paper can be folded two or three times along its length, so that it can be tied to one of the trees in the immediate neighbourhood – a familiar sight in almost any shrine precinct. This allows time for the unfavourable advice to be mitigated, if not reversed. The woodcut print reproduced in figure 6.1 evokes the whole scene very well.

The numbers on the *kuji* each connote one of six basic states, three of which, under the heading *kichi*, 吉, connote good fortune, while three, under the heading *kyō*, 凶, connote ill fortune. In each case the level of fortune is graded *dai*(大), *chū* (中) or *shō* (小), that is 'large', 'middle' or 'small'. In some instances, the *kuji* are not numbered at all, but simply bear the *kanji* representations of the six possibilities, such as, for example, 大吉 or 中凶 . These will then

*Figure 6.1*  Omikuji temple scene

correspond to six rows of the frame, or in mathematical terms, six co-ordinates of the matrix it represents. However the *kuji* are labelled, there is generally a bias towards *kichi* and away from *kyō*, so that the box may contain, say, six *dai-kichi* and only one *dai-kyō* kuji.

The message contained on the slips of paper is generally divided into a number of sections (separated by an 'o' in figure 6.2). Of these, some will mention specific events likely to be significant, if and

*Figure 6.2* Omikuji written advice

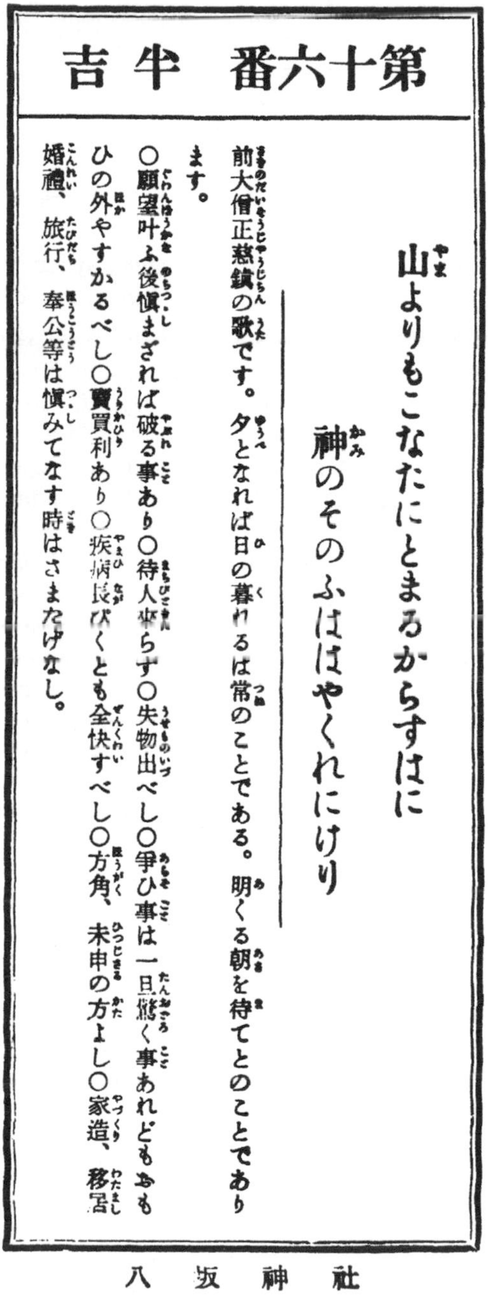

第十六番　半吉

山よりもこなたにとまるからすはに

神のそのふははやくれにけり

前大僧正慈鎮の歌です。　夕となれば日の暮れるは常のことである。　明くる朝を待てとのことであります。

○願望叶ふ後慎まざれば破る事あり○待人来らず○失物出べし○争ひ事は一旦慈く事あれども也○方角、未申の方よし○家造、移居の外やすかるべし○賣買利あり○疾病長びくとも全快すべし

婚禮、旅行、奉公等は慎みてなす時はさまたげなし。

八坂神社

when they occur, while others will note that particular days of the month are auspicious, or inauspicious for specific activities, such as journeying away from home. This is, of course, a numerical factor. The fact that *omikuji* are located in shrines and temples suggest that they are related to the common practice of *genze riyaku*, whereby the Japanese petition their deities and Buddhas for benefits and help in this life. On the other hand, the popularity of *omikuji* with children, which adults do nothing to discourage, suggests that the institution is not always taken seriously.

The word *omikuji* is generally written in the *hiragana* syllabary, but there are a number of recognised *kanji* versions. *Kuji* is the *kun* reading of two complicated *kanji*, 籤 and 圖 , of which the latter is seen as a *zoku-ji*, or alternative reading of the former. *O* and *mi* are honorific prefixes, both represented by the familiar *kanji*, 御. In *o-mi-kuji*, therefore, this *kanji* will be repeated, to form 御御籤, or 御御圖. In the former, but not the latter case, there is an alternative version, 御神籤, where the second *kanji*, 神, has the *kun* reading, *kami*, meaning a 'god' in Shinto.[3]

The use of the double honorific underlines the object of *omikuji* in seeking the will of gods, or of using prophecy (as recorded on the slip of paper) to distinguish between good and evil. *Kuji* standing alone and referring to a material object means a 'skewer', which would be a possible use of the metal rods if they could be taken out of the box containing them. By extension it can mean 'sharp', and that also in a figurative sense. Its numerical property is found in the additional meaning of 'counter', with the alternative of a 'sign' or 'talisman'. That is, *omikuji* provides a numerical sign for telling the will of the gods.

Since this is the essential basis of almost all Japanese fortune-telling, *omikuji* may be seen as its simplest, cheapest and most popular form. This does not mean that *kyūsei* and *ekikyō* are less important, only that they are much more complicated. This being so, they are probably less often consulted. *Omikuji* is divination for every day, with no pretensions for changing the whole course of life.

## KYŪSEI OR THE JAPANESE HOROSCOPE

The Japanese almanac, or *koyomi*, published with every new year in a number of different editions, provides the basis for determining one's fortune, whether good or ill. The categories *kichi* and *kyō* used with *omikuji* apply equally in different situations dealt with in the almanac. This provides for a number of different methods, but everyone of them is essentially numerical. To describe them all within the space of a few paragraphs is next to impossible, but any survey of the use of numbers in Japan would be incomplete without some explanation of the basic principles.

The key to the standard *koyomi* is to be found in the inside cover, next to the first page. This, the *nenrei hayamihyō*, or 'age chart', provides the key for individuals born in every year from 1892 (Meiji 25) to 1990 (Heisei 2), a span of ninety-nine years.[4] The chart, shown in table 6.1 is a matrix of eleven rows and nine columns, to be read from right to left, starting at the top right hand corner. Table 6.2 is a fore-shortened version, representing the *kanji* in table 6.1 by means of an alphabetical code combined with Arabic numerals. The year in the top right hand corner is 1892, with the current year, 1990, in the bottom left. The columns are headed with the numbers from 1 to 9, in descending order when read, as is correct, from right to left. These numbers correspond to the nine stars of the *kyūsei*, or more exactly to nine *kyū*, or 'constellations', listed in table 6.3 (which, as I shall show, provide the arithmetical key to the whole art of *kyūsei*). Each of these also determines the *kishō* or 'temperament' of the year in question, which will correspond to one of the five elements, earth, water, fire, wood and metal. The last number to be read is always that of the current year, which in 1990 happens to be 1 (with the result that, reading from left to right, the columns are simply headed 1, 2, ... up to 9). Since the nine-year cycles are numbered in descending order, 1989 had the number 2 (so that the columns were then headed 2, 3 ... 8, 9, 1).

In their relation to the succession of years the ten *kan* and twelve *shi* do not occur in descending order. In figure 6.2 the former are encoded according to the alternating *yin* and *yang* of the five elements, so that, for instance, M+ is *kanoe*, or the *yang* of metal, where T− is *kinoto*, the *yin* of wood.[5] The twelve *shi* are simply numbered in the order in which they occur in the Chinese zodiac, starting with 1 for 'rat' and ending with 12 for 'boar'.[6] As I note in chapter 7, for the reckoning of time according to the *kanshi*, any combination of cyclical systems defines a new system in which the number of members is equal to the lowest common multiple of the numbers comprised in the component systems. The *nenrei hayamihyō* includes cycles of nine, ten and twelve members: the first two combine to form a ninety-year cycle (which has in fact no particular significance); the last two combine to form the sixty-year cycle of the *kanshi*, and the first and third combine to form a shorter 36-year cycle which has its own place in the almanac. All three systems

*Table 6.1*　Nenrei hayamihyō

| 一白 水気性 | 二黒 土気性 | 三碧 木気性 | 四緑 木気性 | 五黄 土気性 | 六白 金気性 | 七赤 金気性 | 八白 土気性 | 九紫 火気性 | 平成二年 生年・九星・干支 年齢早見表 |
|---|---|---|---|---|---|---|---|---|---|
| 明治・十年生 満九十一歳 | 明治十一年生 満九十二歳 | 明治十二年生 満九十三歳 | 明治十三年生 満九十四歳 | 明治十四年生 満九十五歳 | 明治十五年生 満九十六歳 | 明治十六年生 満九十七歳 | 明治十七年生 満九十八歳 | 明治十八年生 満九十九歳 | （今年の誕生日が来てこの表の満年齢となる。立春以前に誕生の人は前年の九星と干支で見ること。） |
| 明治十九年生 満八十一歳 | 明治二十年生 満八十二歳 | 明治廿一年生 満八十三歳 | 明治廿二年生 満八十四歳 | 明治廿三年生 満八十五歳 | 明治廿四年生 満八十六歳 | 明治廿五年生 満八十七歳 | 明治廿六年生 満八十八歳 | 明治廿七年生 満九十歳 | |
| 明治廿八年生 満七十二歳 | 明治廿九年生 満七十三歳 | 明治三十年生 満七十四歳 | 明治卅一年生 満七十五歳 | 明治卅二年生 満七十六歳 | 明治卅三年生 満七十七歳 | 大正元年生 満七十八歳 | 大正二年生 満七十九歳 | 大正三年生 満八十歳 | |
| 大正四年生 満六十三歳 | 大正五年生 満六十四歳 | 大正六年生 満六十五歳 | 大正七年生 満六十六歳 | 大正八年生 満六十七歳 | 大正九年生 満六十八歳 | 大正十年生 満六十九歳 | 大正十一年生 満七十歳 | 大正十二年生 満七十一歳 | |
| 大正十三年生 満五十四歳 | 大正十四年生 満五十五歳 | 昭和元年生 満五十六歳 | 昭和二年生 満五十七歳 | 昭和三年生 満五十八歳 | 昭和四年生 満五十九歳 | 昭和五年生 満六十歳 | 昭和六年生 満六十一歳 | 昭和七年生 満六十二歳 | |
| 昭和八年生 満四十五歳 | 昭和九年生 満四十六歳 | 昭和十年生 満四十七歳 | 昭和十一年生 満四十八歳 | 昭和十二年生 満四十九歳 | 昭和十三年生 満五十歳 | 昭和十四年生 満五十一歳 | 昭和十五年生 満五十二歳 | 昭和十六年生 満五十三歳 | |
| 昭和十七年生 満三十六歳 | 昭和十八年生 満三十七歳 | 昭和十九年生 満三十八歳 | 昭和二十年生 満三十九歳 | 昭和廿一年生 満四十歳 | 昭和廿二年生 満四十一歳 | 昭和廿三年生 満四十二歳 | 昭和廿四年生 満四十三歳 | 昭和廿五年生 満四十四歳 | |
| 昭和廿六年生 満二十七歳 | 昭和廿七年生 満二十八歳 | 昭和廿八年生 満二十九歳 | 昭和廿九年生 満三十歳 | 昭和三十年生 満三十一歳 | 昭和卅一年生 満三十二歳 | 昭和卅二年生 満三十三歳 | 昭和卅三年生 満三十四歳 | 昭和卅四年生 満三十五歳 | |
| 昭和卅五年生 満十八歳 | 昭和卅六年生 満十九歳 | 昭和卅七年生 満二十歳 | 昭和卅八年生 満二十一歳 | 昭和卅九年生 満二十二歳 | 昭和四十年生 満二十三歳 | 昭和四十一年生 満二十四歳 | 昭和四十二年生 満二十五歳 | 昭和四十三年生 満二十六歳 | |
| 昭和四十四年生 満九歳 | 昭和四十五年生 満十歳 | 昭和四十六年生 満十一歳 | 昭和四十七年生 満十二歳 | 昭和四十八年生 満十三歳 | 昭和四十九年生 満十四歳 | 昭和五十年生 満十五歳 | 昭和五十一年生 満十六歳 | 昭和五十二年生 満十七歳 | |
| 平成元年生 満一歳 | 平成元年生 満一歳 | 平成元年生 満二歳 | 平成元年生 満三歳 | 昭和五十四年生 満四歳 | 昭和五十五年生 満五歳 | 昭和五十六年生 満六歳 | 昭和五十七年生 満七歳 | 昭和五十八年生 満八歳 | |

*Table 6.2* Nenrei hayamihyō (English version)

| 1 | 2 | 3 | 4 | 5 | 6 | 7 | 8 | 9 |
|---|---|---|---|---|---|---|---|---|
| *white* water | *black* earth | *azure* wood | *green* wood | *yellow* earth | *white* metal | *red* metal | *white* earth | *violet* water |
| M33–91 M+/1 | M32–92 E–/12 | M31–93 E+/11 | M30–94 F–/10 | M29–95 F+/9 | M28–96 T–/8 | M27–97 T+/7 | M26–98 W–/6 | M25–99 W+/5 |
| S56–10 M–/10 | S55–11 M+/9 | S54–12 E–/8 | S53–13 E+/7 | S52–14 F–/6 | S51–15 F+/5 | S50–16 T–/4 | S49–17 T+/3 | S48–18 W–/2 |
| H2–1 M+/7 | H1–2 E–/6 | S63–3 E+/5 | S62–4 F–/4 | S61–5 F+/3 | S60–6 T–/2 | S59–7 T+/1 | S58–8 W–/12 | S57–9 W+/11 |

*Table 6.3* The nine kyū

| No. | kanji | kyū | constellation |
|-----|-------|-----|---------------|
| 1 | 坎 | *kan* | *konnan* |
| 2 | 坤 | *kon* | *chūi, jibi* |
| 3 | 震 | *shin* | *kaiun* |
| 4 | 巽 | *son* | *fuku-un* |
| 5 | 中 | *chū* | *seikan* |
| 6 | 乾 | *ken* | *kyōun* |
| 7 | 兌 | *da* | *kiraku* |
| 8 | 艮 | *gon* | *henka* |
| 9 | 離 | *ri* | *chōjō* |

combined would create a 180-year cycle. Since, however, this system
has no significance of its own, it does not matter that it is beyond the
99-year range of the *nenrei hayamahyō*.

The chart also records the official name of every year, using the
abbreviations M (Meiji 1868–1911),[7] T (Taishō 1912–25), S (Shōwa
1926–88) and H (Heisei 1989–  ), followed by the number of years
which have elapsed up to and including the present year (1990). This
makes it simple for the user, who will certainly know how old he is
and the year in which he was born, to find out his *kyū, kan* and *shi.*
This is important for using the tables in which both the *kyū* and the
*shi,* may be used for determining affinity between a man and a
woman, as it is in the current year. The use of the former is illustrated
by table 6.4, which is an English version of a table taken from a
standard Japanese almanac for the year 1983.[8]

An example will illustrate the use of this table. A 32-year-old man,
born in 1958, will want to know whether it would be auspicious to
marry a 25-year-old woman, born in 1965. From the *nenrei hayamihyō* he will see that his *kyū* year is 6, and hers, 8, and the chart will
show that this is *daikichi,* or highly auspicious. This factor is determined not so much by the number of the *kyū* year, as by the corresponding *kishō,* which is next to it in the table.[9] The man could go
further and look at the affinity according to the corresponding table
for the twelve *shi,*[10] which would show him as *inu,* or 'dog', and the
lady, as *mi,* or 'snake'.[11] Here he would be less fortunate, since the

*Table 6.4* Affinity between man and woman

| kyū year/ kishō | MAN | | WOMAN | |
|---|---|---|---|---|
| | Daikichi | Chūkichi | Daikichi | Chūkichi |
| 1 W | 6, 7 | 1, 3, 4 | 3, 4 | 6, 7, 1 |
| 2 E | 9 | 5, 8, 6, 7, 2 | 6, 7 | 2, 5, 8, 9 |
| 3 T | 1 | 9, 4, 3 | 9 | 1, 4, 3 |
| 4 T | 1 | 9, 3, 4 | 9 | 1, 3, 4 |
| 5 E | 9 | 6, 7, 8, 2, 5 | 6, 7 | 2, 8, 9, 5 |
| 6 M | 2, 5, 8 | 1, 7, 6 | 1 | 2, 5, 8, 7, 6 |
| 7 M | 2, 5, 8 | 1, 6, 7 | 1 | 2, 5, 8, 6, 7 |
| 8 E | 9 | 2, 5, 6, 7, 8 | 6, 7 | 2, 5, 9, 8 |
| 9 F | 3, 4 | 2, 5, 8, 9 | 2, 5, 8 | 3, 4, 9 |

table gives no affinity in this case. The almanac gives only indirect guidance as to whether more attention should be paid to *kishō* or *shi* affinity; this is one of the many cases in which the advice of an *ekisha*, or professional fortune-teller, is sought.

Another variation on the same theme is simply to use the sixty *kanshi* compounds as the basis for attributing character according to the year of birth. Here the twelve *shi* were seen as particularly important, so that those born in the year of the rat (*ne*) were restless, in contrast to those born in that of the ox (*ushi*) which followed, who were expected to be patient. The most notorious combination is *hinoe uma* (fire-yang/horse); women born in this, the forty-third year in the *kanshi* cycle, are seen as headstrong and liable to kill their husbands. It is said that in 1966, the last such year, the Japanese birth-rate was exceptionally low.

The questions so far considered are only incidental to the main use of the almanac, which is to determine the interests of an individual, from day to day throughout the year, on the basis of the *kyū* of the year of his birth.[12] The key in all such cases is the *hōiban* or 'direction board', represented in figure 6.3. This is in the form of a *mandala* representing the number 9, or in binary form 1001. The core (*manda*) is inscribed *chūō*, which simply means 'centre'. There are three separate containers, or *ra*, which, although all reproduced in the figure, must be seen as alternatives. The components of each are defined in relation to the eight cardinal points, N, NE, E, SE, S, SW, W, NW, seen from the north, or bottom of the page, where the observer stands, looking south.[13] In the first container the points correspond to the eight basic *hakke*, or trigrams, of the *I Ching*,

*Figure 6.3* The basic hōiban

# 方位盤の図

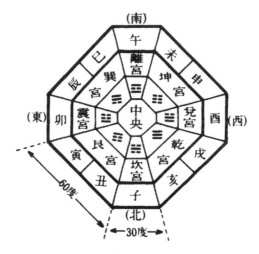

which I describe in the following section. In the second container, together with core, are to be found the nine *kyū* (listed in table 6.3) which are the basis of *kyūsei*. In the third container are the twelve *shi*, but so distributed that there are two in each of the intermediate points, NE, SE, SW and NW.

The important point is the order of the nine *kyū* in the second container, which is not that of the numerical order in table 6.3, in contrast to the twelve *shi* in the third container, which are in the correct order. The answer is to be found in the magic square represented in figure 6.4, which according to legend was found inscribed on the back of a sacred tortoise which once upon a time emerged from the water in Kyoto (Yoshino 1983: 53).[14] If then the *kanji* in the *hōiban* are replaced by the corresponding numbers from table 6.3, their order will be the same as that in figure 6.4.

The stage has now been reached of being able to construct the basic scheme of the almanac. The basis is the nine years of the *kyūsei* cycle, with a separate calendar for every single month in the current year for determining the fortunes of individuals born in any one year of the nine comprised in a full cycle. Every such year is governed by a *hōiban* based on that of the current year, which can be read from figure 6.3. The key is always the number at the centre of the square,

*Figure 6.4* The directional magic square

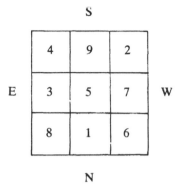

S

|   |   |   |
|---|---|---|
| 4 | 9 | 2 |
| 3 | 5 | 7 |
| 8 | 1 | 6 |

E (left)   W (right)

N

and because 1990 has the number 1 in the *kyūsei* cycle, the top left-hand square in figure 6.5 is the model for the *hōiban*.[15] Because the *kyūsei* cycle of years operates in *descending* order, 1989 was governed by the square with 2 at its centre, 1988 by that with 3, and so on down to 1981, when the cycle completes a full turn with the same *hōiban* as 1990. Future years will be governed by the same principle, so that 1991 will correspond to 9, 1992 to 8, with the next 1-cycle being that of 1999.

For any given year, the nine yearly calendars govern the fortunes of individuals born in any year in the *kyūsei* cycle. To take the example of someone born in 1929, the *nenrei hayamihyō*, shows this year to have the number 8. The fortune of such an individual must then be sought in the eighth yearly calendar for the year 1990. This will be governed by the 1990 *hōiban* (based on 1 in the *kyūsei* cycle) but with its own particular distribution of good and ill fortune, *kichi* and *kyō*, round the eight cardinal points, determined by its relating to a person born in the eighth year of a *kyūsei* cycle.

Each month in the eighth yearly calendar will also have its own distribution of *kichi* and *kyō*, imposed on the 1990 (1-*kyū*) *hōiban*, but the actual *kyūsei* cycle of days is defined for the whole year, and in contrast to the *kanshi* cycles, is the same for every year. The reason is that the *kyūsei* cycle follows the sun, so that it ascends in the order 1, 2, ... up to 9 in the period starting with the winter solstice (which is taken to be 24 December), and descends in the order 9, 8 ... down to 1 in that period starting with the summer solstice (which is taken to be June 27).[16] Logically enough, the summer turning point is 9,[17] and

*Figure 6.5* The order of numbers in the nine hōiban in the kyūsei cycle

| 9 | 5 | 7 |
|---|---|---|
| 8 | 1 | 3 |
| 4 | 6 | 2 |

| 1 | 6 | 8 |
|---|---|---|
| 9 | 2 | 4 |
| 5 | 7 | 3 |

| 2 | 7 | 9 |
|---|---|---|
| 1 | 3 | 5 |
| 6 | 8 | 4 |

| 3 | 8 | 1 |
|---|---|---|
| 2 | 4 | 6 |
| 7 | 9 | 5 |

| 4 | 9 | 2 |
|---|---|---|
| 3 | 5 | 7 |
| 8 | 1 | 6 |

| 5 | 1 | 3 |
|---|---|---|
| 4 | 6 | 8 |
| 9 | 2 | 7 |

| 6 | 2 | 4 |
|---|---|---|
| 5 | 7 | 9 |
| 1 | 3 | 8 |

| 7 | 3 | 5 |
|---|---|---|
| 6 | 8 | 1 |
| 2 | 4 | 9 |

| 8 | 4 | 6 |
|---|---|---|
| 7 | 9 | 2 |
| 3 | 5 | 1 |

the winter turning point, 1, both numbers being repeated. Since the number of days in the year is never divisible by 9, a break must always occur, and this is now made when the old year turns into the new. The *kanshi* cycles, like the days of the week, continue uninterrupted from one year to the next, so that in this case every year will be different.[18]

For every day, in every month, in each of the nine years, the *kyūsei* cycle defines the appropriate level of fortune, ranging from *daikichi* to *daikyō*, on the same principle as that governing the *omikuji*. In this case, however, there are only five categories, *daikichi* (++), *kichi* (+), *hanhan* (±), *kyō* (−) and *daikyō* (−−), which occur according to figure 6.6. There is little rhyme or reason to the distribution of the signs in this figure save that *hanhan*, literally 'half and half', always occurs when the number of the day in the nine-day cycle is the same as that of the year in the nine-year cycle. At the same time, all five signs occur with roughly the same frequency.

The basis for the fortune of every individual, on every single day of the year, is now complete. To give an example, for someone born in 1929 (*kyūsei* 8), 10 June in 1990 (*kyūsei* 1) is day 1, and so, from figure 6.6, counts as *kichi* (+), or mildly auspicious.[19] Figure 6.6 is not, however, part of the almanac, although, without any doubt, something like it must have been used in compiling the almanac. The actual user simply looks up 10 June, in the cycle of nine-yearly calendars for the year 1990. In doing so he will learn not only his *kikkyō* rating (+), but also find advice written specifically for this day, for someone born in a *kyūsei*-8 year, such as 1929. In the example chosen the advice reads:

Readiness to be absorbed in setting up an objective is decisive.
Lack of sleep is the root cause of injury to health

which of course is precisely what the typical Japanese workaholic ought to hear. If the moral is that one must not work too hard, and at the same time make certain of sufficient sleep, then my analysis is certainly a roundabout way of reaching this point. But this is not the point at all. The Japanese who consults an almanac has a quite different object from that of the non-Japanese scientist who is concerned to explain the principles upon which the almanac is compiled. In the end these have proved to be surprisingly simple. The key is numerical. It is to be found in arranging the nine numbers of the *kyūsei* cycle to form a magic square, which is then the basis of a cycle of nine *hōiban*, as illustrated in figure 6.5, while at the same time applying the same nine numbers, alternating at the solstices between ascending and descending order, to the yearly calendars, to produce the *kikkyō* ratings in figure 6.6.

The analysis given above, seen at first sight, may look like using a heavy sledge-hammer to crack a very small nut.

## THE EKIKYŌ OR BOOK OF CHANGES

In the history of fortune-telling no single work has been more influential than the *Book of Changes*, generally known by its familiar

*Figure 6.6*  Kyûsei cycles of good and ill fortune

| 9-day cycle \ 9-year cycle | 1 | 2 | 3 | 4 | 5 | 6 | 7 | 8 | 9 |
|---|---|---|---|---|---|---|---|---|---|
| 1 | ± | + + | + | − | + | − − |  | + | + + |
| 2 | + + | ± | + + | + | − | + | − − | − − | + |
| 3 | − − | + + | ± | + + | + | − | + | − − |  |
| 4 | ± | + | − | ± | + + | + | − | + | − − |
| 5 | − − | − | + | + + | ± | + + | + | − | + |
| 6 | + | − − | ± | + | + + | ± | − − | + | − |
| 7 | − | + | − − | ± | + | + + | ± | + + | − − |
| 8 | + | − − | + | − − | − | + | + + | ± | + + |
| 9 | + + | + | − | − − | + | − | − − | + + | ± |

Chinese name, *I Ching*. The Japanese version, known as the *ekikyō*, is seen as the foundation of all divination, and diviners are generally known as *ekisha*, rather than as a simple practitioner of *uranai*, the word for divination introduced at the beginning of this chapter. In spite of its great antiquity, the *ekikyō*, even in the present century, has been consulted at critical moments in Japanese history. The Japanese high command had little hesitation in attributing the astounding victories won at the beginning of the Pacific War to books of strategy based upon the *ekikyō*, which were required reading for senior officers (Blofeld 1976: 15). The actual decision to attack Pearl Harbor on 8 December, 1941 was also taken only after an *ekisha* had been consulted.

The *ekikyō* is essentially an oracle. This means that the answers it gives are open to alternative interpretations, so that once the tide turned against the Japanese in the Pacific War, it was not the *ekikyō* but the manner in which it was used which was at fault. At the same time the basis of the *ekikyō* is numerical. The greater part of it consists of explanatory texts for sixty-four signs, known in English as hexagrams, each consisting of six lines, which may be broken or unbroken. Each hexagram is a combination of two out of eight possible trigrams, or *hakke*, each consisting of three lines.[20] Both the trigram and the hexagram are read from the bottom line up. The basic relationships in binary mathematics are quite transparent: $8 = 2^3$, $64 = 8^2 = 2^6$. The eight trigrams are given in table 3.1 (see p. 37), together with their names and corresponding *kanji*, and these, as I have already noted, are the same as those of the eight *kyū* surrounding the core of the *hōiban* in figure 6.3. The names also have a *meaning*, for which there is, in every case, an ordinary Japanese *word*. These are also given in table 3.1. The fact that the order is different from that in table 6.3 is not significant, since there is no one fixed order for the *hakke*.[21] The important point, already noted, is that the two systems are inter-locked.

In Chinese each of the sixty-four hexagrams has its own name, with its own meaning, which can be taken to indicate both the correct interpretation and the contexts in which it is applicable. Where the two component trigrams are identical, their name is also that of the hexagram. In Japanese the name in such a case, on the basis of table 3.1, takes the form 'name-*i*-word', so that the first hexagram becomes *ken-i-ten*, literally 'ken *becomes* ten'. This rule applies only to six cases out of sixty-four; the remaining cases are named according to the two component trigrams, followed by the Japanese version of the Chinese name of the hexagram. Taking as an example 遯 , the Chinese name is simply *tun*, with *ton* as the Japanese equivalent. The full Japanese name then becomes *ten-zan-ton*, which, since it indicates the two component trigrams, is more informative than the simple Chinese *tun*.

For every hexagram there are (in addition to the two *component* trigrams, constituted, respectively, out of rows 1, 2 and 3, and 4, 5 and 6) two *nuclear* trigrams, constituted out of rows 2, 3 and 4, and 3, 4 and 5. According to the particular case, both the component and nuclear trigrams can contribute to the meaning of the hexagram, although only the former are part of its name. In addition every hexagram has either one or two *ruling* lines, which according to tradition, are particularly important in determining its meaning.[22] Finally, the answer to any question depends not on one but on two hexagrams, from which the second derives, and qualifies the meaning of the first. It is this transformation which lies at the root of the *change* in the title of the book.

The correct use of the *ekikyō* is subject to ritual prescribed by tradition. Although, in the context of numbers, only that part of the ritual is essential which determines which two hexagrams provide the answer to the question put to the oracle, the whole ritual is regarded as critical. For its performance a number of objects are required in addition to the actual book. These are an incense-burner, a tray, writing materials and a box, with a lid, containing fifty divining sticks, together with a table upon which to place all the objects to be used. The *ekikyō*, wrapped in a special cloth, and the divining sticks, in their box, are kept together on a special shelf.

The subject sits at the table with his back to the south, so that the actual book faces south, as a sign of its authority.[23] The question to be asked has already been framed, and at every stage in the ritual the subject (S) must concentrate upon it. He takes up his position at the table, lights the incense-burner, takes the divining sticks out of the box with his right hand, and passes them three times through the smoke of the incense. One stick is then returned to the box, leaving the remaining forty-nine to be used in determining the form of the hexagrams. According to tradition, every line, starting at the bottom, is determined separately, so that the procedure is performed six times in all. Its basis is always the same: the forty-nine sticks are divided quite at random, into two heaps, one on the right and one on the left. The procedure comprises a number of steps:

1  S places one stick from the right heap between the ring and little finger of his left hand.
2  S then reduces the left heap four sticks at a time until there are only one, two, three or four left. This remainder he places between the middle and ring finger of his left hand.
3  S repeats step 2 with the right heap, placing the remainder of one, two, three or four between the index and middle finger of his left hand.
4  The sticks held in the left hand are counted, and put on one side.
5  With the remaining sticks the whole procedure is twice repeated, so that in all S will three times count sticks held in his left hand,

obtaining as a result a combination of three numbers. This then determines the form of the line according to the table 6.5.

At first sight the results given in the table are incomplete, but this is not so. For arithmetical reasons (based on an elementary instance of the mathematical theory of congruences) 5 and 9 are the only possible results of the first count, where 4 and 8 are the only possible results of the second. The table is therefore complete; there are no other possibilities.

The procedure may seem complicated, but an experienced user can probably complete it in under a minute, with the same sort of dexterity as is to be found with the skilled abacus users described in chapter 9. As every line is determined, according to its type, it is written twice. If it is an *old* line, it will change with the second time; if *young*, it will remain the same.

The final result is two hexagrams side by side, which are then interpreted according to the rubric contained in the *ekikyō*. Here every one of the sixty-four hexagrams is presented in the same way. The basic meaning of the whole hexagram is given in a short and generally cryptic text of two or three lines.[24] This is amplified by a rather longer commentary, which is not always transparent. The symbolism of the hexagram is then explained according to the meaning of the two component trigrams, as given in table 3.1. There is then a shorter text and commentary for each of the six separate lines, but this is only important when they are *old* or *moving*, lines.

Applying the rule of change, the first hexagram answers the question put by the subject according to the present state of affairs,

*Table 6.5* The lines of trigrams and hexagrams

| Combination of sticks | Total | Form of line | Type of line | Ritual number |
|---|---|---|---|---|
| 5, 4, 4 | 13 | —— o —— | old yang | 9 |
| 9, 8, 8 | 25 | —— x —— | old yin | 6 |
| 5, 8, 8<br>9, 8, 4<br>9, 4, 8 | 21 | ———————— | young yang | 7 |
| 5, 4, 8<br>5, 8, 4<br>5, 4, 8 | 17 | ——   —— | young yin | 8 |

where the second suggests future developments (which are often important for determining the action to be taken). The moving lines, interpreted according to the rubric of the first hexagram, indicate the particular matters to be born in mind in making the transition.

Cosmic order is the underlying principle of the *ekikyō*. The hexagrams, in their proper order, provide a model which can guide the individual when important decisions, such as whether to accept new employment, have to be made. The moral basis for making the decision is harmony with the course of the universe, in which every individual has a certain destiny. The practical problem is that the answers given by the hexagrams are only too often cryptic, equivocal and even self-contradictory.

As in interpreting any oracle, intuition plays a key part, and for this reason much importance is attached to the enquirer's state of mind. If, in the context of number, the manipulation of the divining sticks seems the most significant part of the ritual, an experienced practitioner could well see this as a purely mechanical operation. What counts is his own state of mind, as shown in the reverence with which he handles the sacred objects used in the ritual and interprets the results which they then yield.

## THE METAPHORICAL CONUNDRUM

To any scientific observer, the successful use of numbers in divination must require that they establish some sort of metaphor. In other words, numerical divination, in the way in which it governs decisions in daily life, must be an alternative to econometrics, seen as the *science* by which numerical models are constructed for determining economic policy. The way the numbers behave in the model is a refraction of the manner of operation of the sector of the economy from which it is derived. In principle, every new model can be put to the proof, even though in practice most such models are destined to be no more than the currency used by academic economists in dealing with each other. The principle is none the less adhered to: it would be the grossest heresy to deny the possibility of constructing a useful econometric model of any actual economic situation.

The use of a model for regulating economic activity is quite general, even in cultures in which any formal science of economics is quite unknown. In such cases, as Gudeman (1986) shows, the actual model used is seen as a refraction, not of an economic system, but of a social order. So long as the social order, regulated according to the model, supports a viable economy, the model cannot be said to fail. It may be that every department of life is regulated by metaphor; all that is necessary is that, sooner or later, the counter-productive metaphors be discarded. The only problem is that the use of metaphor is not so much a question of application as of interpretation. This means that defects are human, not faults in the system. Indeed a purely numerical system has no tolerance for faults at all. The mathematics underlying the *kyūsei* or *ekikyō* may be elementary, or even trivial, but it contains no mistakes. As a model, it is perfect, and this is precisely its appeal. The point is appreciated also in Japan, where, according to Yoshino, 'divination (*eki*) requires a unified logic of the universe to be found by conceiving of the reduction of form to number' (Yoshino 1983: 57).[25] The numerical system is a model of the universe, of which the *kyūsei* and *ekikyō*, and even the elementary *omikuji*, are refractions.

The root problem lies in the implicit assumption that the numbers, which may occur in encoded form, such as the hexagrams of the *ekikyō*, are metonymic, that is, that they are not only *names* for, but also incorporate some *attribute or cause or effect*[26] of the associated texts. The implicit basis of all numerical divination is a metaphor in which the effective components are metonymies. If the metonymic attribution fails, then the whole foundation collapses. Judged in this light, the whole system, however ancient, is fundamentally flawed.

This is not the end to the matter. The moral is not necessarily that some econometric model should be preferred, just because its metonymic connections are not spurious.[27] Such models fail because of defects in the quality of the information on which they are based. The divination model, with its cosmic pretensions, does not even claim to be based on information of this kind. This is why intuition is essential in using it, and with reasonable goodwill there is every possibility that intuition corresponds to common sense. This was the breakdown in the system when the Japanese high command launched its attack on Pearl Harbor.[28]

# 7 Time

## THE JAPANESE VOCABULARY OF TIME

Every culture has to wrestle with the problem of dealing with the concept of time. Time is seen as something to which we are all subject, but which can and indeed must be regulated, if not mastered. Leach suggests that all aspects of time are derived from 'two basic experiences: (1) that certain phenomena of nature repeat themselves and (2) that life change is irreversible (Leach 1966: 125).

Leach goes on to assert that religion constantly requires the two basic experiences to be embraced under one category:

> Repetitive and non-repetitive events are not, after all, logically the same. We treat them both as aspects of one 'thing', *time,* not because it is rational to do so, but because of religious prejudice. The idea of Time, like the idea of God, is one of those categories which we find necessary because we are social animals rather than because of anything empirical in our objective experience of the world. (ibid.)

Language is the key to the concept of time in any culture. Numbers may then provide the means for using this concept as a means for organising activity in everyday life. Before dealing with numbers in relation to time in Japan (which is the main theme of this chapter) it is necessary to see how time appears in the Japanese language. The *kanji* used for the different words in the written language can also be analysed so as to add to our understanding of the way the Japanese conceive of time.

The basic character is 時, with an *on* reading, *ji* and a *kun* reading, *toki,* which is the 'only pure Japanese word for time' (Caillet 1986: 45). Although the meaning of *toki* is quite general, it occurs particularly in contexts where in English it would be translated by 'when', so as to indicate the time at which something specific happens, or is expected to happen.[1] In other words, *toki* is the time of an *event*. The radical, 日, of the character, 時, generally in its *kun* reading, *hi*, means 'sun'[2] or 'day'. The *on* reading of 時, *ji*, occurs in any number of compounds, all of which connote time in some way. Of these, four in particular show how the Japanese conceive of time: *jiten*, *jikoku*, *jikan* and *jidai*.

*Jiten*, in *kanji* 時点, means 'a *point* in time', in principle an instant of time without duration. In this it corresponds to words in several western languages, such as the Dutch *tijdstip*. It connotes much of what is summed up in the English 'punctual', that is that a point in time can be identified in advance and specified as the instant at which something is to take place.

*Jikoku*, in *kanji* 時刻, has much the same meaning, although it is used in different contexts, such as *jikokuhyō*, the normal word for a time-table. The *kun* reading of the same *kanji* is the verb *kiza(mu)*, which can mean to 'cut' or 'notch', so that *jikoku* imports the recording of time in this way. The use of the word in *jikokuhyō* is therefore appropriate, for it relates the dimension of time to that of space.

*Jikan*, in *kanji* 時間, is a word connoting duration, not the instant moment. *Kan* is the interval[3] between two *ten*,[4] in time or space, so that *jikan* is the normal word for 'hour' although it can mean simply 'time'.[5] The normal word for an interval is the *kun* reading, *aida*,[6] equivalent to *kan*, which refers particularly to the time between the days, *setsu-bi*, for recognised rites. *Setsu* occurs also in the normal word for 'season', *kisetsu*,[7] and its original connotation was that of a '(bamboo) joint' – by implication between two *aida* (Caillet 1986: 34).

*Jidai*, in *kanji* 時代, is perhaps the most significant of the four *ji*-compounds to be analysed. This word, familiar to any Japanese, means 'period' or 'era', which would also be the meaning of *dai* if it stood alone. The *kun* equivalents to *dai*, in the verbal forms *ka(eru)* and *ka(waru)*, import 'substitution', 'renewal', 'change' and 'alternation'. The commonest use of *jidai* is probably to designate an era corresponding to the reign of an emperor, so that one now talks of the *Shōwa-jidai*. *Jidai* therefore implies succession, and with succession, a new *zeitgeist* for every era. This point will come up again with the detailed examination of the Japanese calendar.

Besides *ji*, with all its compounds relating to time, two other *kanji*, 基 and 機, commonly connote time. Both have the *on* reading, *ki*, but this is pure coincidence. The former, 基, is particularly appropriate for statements relating to future time. An example is *kitai*, meaning 'hope' or 'expectation', in which the component, *tai*, means 'waiting'. Many compounds imply the limitation, also, of time, so that *kigen* is a time limit, or deadline, and *kikan* is a period with a definite starting point, *kishu*, and endpoint, *kimatsu*. (This clearly distinguishes it from *jidai*, which is a period in the sense of an *era*.)

The connotations of *ki*, written with the *kanji*, 機, are in a quite different semantic area. Not surprisingly, for a *kanji* whose *kun* reading, *hata*, means 'loom', most of the compounds connote something mechanical, but *kikai* is a quite normal word for 'chance' or 'opportunity', and other compounds, such as *kigi*, *kien* and *kiun*, have much the same meaning. Here again time is evoked in terms of the alternation, with *kikai* being determined by the place where, instantaneously, the shuttle stops. The connotation is relatively weak in this case, and its metaphorical basis comparatively modern.

In words relating to time, the connotation of cyclic movement is much weaker than that of alternation. Common *kanji* such as 回 and 周, connoting circularity in most of their compounds, do not relate to time. 回, which combines in its *on* reading, *kai*, with numerals to mean '-times' (so that, for instance, *nikai* means 'twice'), implies circularity in almost every variant of its *kun* readings, so that *mawa(ri)* can mean 'rotation' or 'circumference'. *Mawari* is also a *kun* reading of 周, which in its *on* reading, *shū*, has any number of compounds implying a recurring cycle – so much so that *shūha* and *shūki*[8] are both normal words for 'cycle' or 'frequency' in the wave theory of physics. The *kanji* 周 also provides the *phonetic* component of 週, whose only meaning is 'week', almost always in the compound *shūkan*.[9] The question is, how significant is it that a measured period of time, that is, the week, which was only introduced into Japan in the modern era (Chamberlain 1974: 475), is the one with the most pronounced *cyclical* basis? An answer must wait on the analysis of the cyclical nature of the traditional calendar.

Finally, it is worth noting that the two periods of time, defined in terms of the suffix *kan*, that is, *jikan* and *shūkan*, both have no basis in natural, or better, celestial phenomena. The *hour*, as a fixed part of the day, is a cultural invention, just as is the *week*, as a period of a fixed number of days. In traditional Japan, because the length of the day was determined by the period of daylight, the length of the hour varied according to the season, whereas the length of the week remained, *a fortiori*, constant. The hour also, as it is divided into minutes, or *jibun*, makes use of a unit, derived from the Chinese *fen*, which means the smallest unit into which anything can be divided. The corresponding *kanji*, 分, with, in Japanese, the *on* reading, *bun*, has a *kun* reading, *wa(keru)*, meaning 'to divide'.[10]

## TIME AND NUMBER IN JAPAN

In the first instance time is experienced as a succession of events, external to the observer, and occurring as part of the order of the cosmos. These events are defined by the apparent movement of the sun and moon. At the most elementary level this is seen as mere alternation, the sun by day is followed by the moon by night – an endless oscillation between *yang* and *yin*. Although theoretically this could provide the basis for relating number to time, a binary system of this kind would not only be operationally inconvenient, but would also fail to relate to any other observations of the sun and moon. In practice, if numbers are to be applied to time, the basis must be some period defined in terms either of the observation of the sun and moon, or of the experience of natural phenomena directly linked to their movement across the skies.

On this basis the starting point in the use of numbers must be the counting of days. This leads immediately to a problem which arises whenever it comes to counting any unit of time. In the general case of objects to be counted, the process of counting is controlled by the individual, who chooses not only the order but also the pace at which the process is carried out. This is inherent in the *vicariant* order of the members of a *denumerable* set, as presented in chapter 2. It is an inherent property of time that the units to be counted determine their own pace and order. Both these factors need to be considered further.

The normal process of counting occurs at a pace determined by the individual. Time, in contrast, determines its own pace, which, on any occasion of ordinary daily life, is so slow that some means must be adopted to help in the counting process. The means may be mechanical, as with a clock, but this plays little role in any long-established tradition. The counting of days, let alone months or years, requires some means of recording time systematically. This is the function of the calendar, which I consider in the following section.

The essential *ordinal* property means that when time is counted the object is to establish an order of defined events. This may be entirely prosaic, so that Fukuzawa (1966: 60) notes how in his student days in the last years of the Tokugawa shoguns the food provided was determined by the last number occurring in the day of the month,[11] rather as a western institution might vary its menu according to the day of the week. This case is one of a cycle which repeats itself every month, but the ordering of days – or years – in the life of an individual is equally significant.

In Japan this is to be observed both in the ceremonies following the birth of a new-born child, which occur at fixed numbers of days from the date of birth,[12] and in the domestic rites performed after a death. The former belong to Shinto, the latter, which are much more protracted, to Buddhism. There is an explicit parallel between the two, relating the numbers of the days following a birth, to the numbers of the years following a death. In the case of a birth the series of rites culminates in the presentation of the new-born child at the local shrine, is a ceremony known as *hatsu-miyamairi*,[13] or 'first shrine visit'. This takes place on the thirty-second day for a boy, and on the thirty-third for a girl. In the case of a death the final rite, the *tomurai-age*, takes place on the thirty-third, or sometimes even the fiftieth anniversary. With death also there are domestic rites prescribed to take place after the lapse of a fixed number of days, of which the *shonanuka* (initial seven) occurs on the seventh day, the *chūin* (centre yin) on the forty-ninth day, and the *hyakkanichi* (hundred days) on the hundredth day. During the period in which these rites are performed, the spirit of the deceased is known as *shirei* and is regarded as a potential cause of harm (Smith 1974: 102). The period is terminated on the first *bon*,[14] the Buddhist festival which every 13–15 August commemorates departed ancestors.[15] These are then joined on 16 August, by the *shirei*, transformed into harmless *shōrō*, who are sent off in a special ceremony, the *shōrō okuri*.

Numerically *bon* is significant for the transition from a period of counting days, to one of counting years. Now, although in certain circumstances days are counted to quite a high number – so that for instance the period of storms is taken to begin on *nihyaku tōka*, the 210th day of the year, and to end on *nihyaku hatsuka*, the 220th day[16] – beyond a certain point, time must be reckoned in terms of months and years. At this point, then, the transition is also from a period of time, the day, experienced in terms of alternation, to the month and the year, which are essentially cyclical phenomena.

The numerical importance of this transition cannot be gainsaid, since some point in time can always be chosen as both the beginning and end of the cycle. Every month repeats the death of the old moon and the birth of the new, and so provides a definite point at which the counting process can begin again. Then, every new month can also be counted, until once again a point is reached at which the succession repeats itself. There is, in this second case, a very important distinction. The astronomical event is defined not by the precise instrumentality of the moon, as it is observed in the night sky, but by epiphenomena, caused by the sun, such as the length of the day in the dimension of time or the length of a shadow in that of space. These cases require a measuring instrument, whether it be a clock or a ruler. If the problem of measurement was solved by the Chinese long before the first contact had been established with Japan (so that there was never any doubt about when such events as the solstices and the equinoxes occurred), such learning never inclined pre-modern Japan to abandon the phases of the moon as the primary means of measuring the passage of time.

It follows then that even if the actual experience of the year was determined by the cycle of the seasons, and the natural changes which accompanied it, the year was still defined in terms of twelve lunar months, each with twenty-nine or thirty days. The new year then began on the first day of the first lunar month, which actually occurred at about the beginning of February, when the festival of *setsubun* still marks the end of winter. From the very beginning, experience had shown that such regulation was anomalous, simply because twelve lunar months fall some twelve days short of a complete solar year. The Chinese solution was simply to correct this discrepancy by providing for a thirteenth, intercalary month, approximately every two and a half years,[17] the correction being based upon the occurrence of the winter solstice.[18] The result is that for Japan, as much for China, two recurrent cycles occur in the experience of time, the first defined by the days of the month, the second by the months of the year. In both cases, by definition, the process of counting begins again at the end of every cycle.

If the days of the month, and the months of the year, are almost by definition repetitive, what about the succession of years, upon which the movement of the heavenly bodies imposes no obviously repetitive order?[19] Does not this provide the context in which, for every individual, in the words of Leach cited above, 'life change is irreversible'? And if so, has every year in such a life the same meaning?

The Japanese are in little doubt about the answers to such questions. First for children, there is the joyful festival of *shichi go san*, celebrated on 15 November in every year. Boys of 5, and girls of 3 and 7, are then presented at the local shrine to ask for the protection of the family gods, or *ujigami*. The period of childhood comes to an end in the twentieth year of life, when coming of age is celebrated on 15 January, a national holiday known as *seijin no hi*. There is no fixed form for the ritual celebration of this predominantly secular event, but visits to the local shrine are common, and in Kyoto there is an archery contest at the Buddhist temple of *Sanjūsangendō*.[20] The rites of childhood come to a definitive end with marriage. For the adult, married and concerned to bring up a family, certain years, known as *yakudoshi*, are much more fateful.

The *yakudoshi* are certain years in the life of the individual in which he is considered to be particularly subject to illness or misfortune. Since, in counting the years, the calendar year in which the individual is born is numbered 1, the *yakudoshi* begins with the year before that in which the actual age is reached. Every connotation of *yaku*, when written with the *kanji* 厄, suggests trouble or ill fortune, and it is characteristic of the Japanese that certain days, *yakubi*, and even more, certain years, *yakudoshi*, should be approached with a sense of foreboding. In the *yakumae*, the year preceding a *yakudoshi*, the individual should do his best to take precautions against his becoming *yakumake*, that is struck by disaster. Although the origins of the belief in *yakudoshi* are obscure (Lewis 1986: 166), Shinto nows offers the means for counter-action. At many a shrine a large notice board lists the principal *yakudoshi*, for both men and women, at the same time offering the services of a priest for the necessary counter-measures (figure 7.1).[21] The more economical can be content with the purchase of a charm (*fuda*) or talisman (*mamori*) (ibid.: 168).

So many years have been identified as *yakudoshi* that it is difficult to establish any sort of order. The greatest threat is to a man of 42 and a woman of 33, but the ages of 25 and 61 are also dangerous for a man, and 19 and 37 for a woman; in many other years also precautions must be taken.[22] No obvious principle governs the choice of numbers for *yakudoshi*. Some have suggested explanations based upon homonyms, others, mathematical formulas. Examples of the former are 42, where 4 – 2 can be read *shi ni*, meaning 'to death', or 33, where the reading *sanzan* can mean 'difficult' or 'troublesome'.[23] The explanation for 61 is plainly numerical, because this is the year of *kanreki*, or returning to the traditional Chinese calendar after the completion of a full sixty-year cycle. It is significant that no *yakudoshi* occurs after the sixty-second year. On the contrary, such special years as occur thereafter, such as the 77th, 88th and 99th, are auspicious for those who are fortunate enough to reach them.

The succession of years recognised as *yakudoshi* belong to the non-repetitive events which show that 'life change is irreversible'. The Japanese see them as obstacles to be overcome in a journey which

*Figure 7.1* Yakudoshi

proceeds only in one direction, ending inevitably with death.[24] The Japanese, defying logic as Leach (1966: 125) sees it, have elaborated the means of incorporating this progress into the recurring cycle of nature, so that the life-cycle of the individual, and the annual cycle of the seasons which rules the life of the community to which he belongs, are equated with each other.

The principles governing this equation, and the means by which it is achieved are involved, are not always consistent. Smith sees Japan as 'a society presenting us with a host of phenomena that at once resist tidy exposition and appear always to lie just outside ready comparative frameworks' (Smith 1974: 214). In particular, 'the Japanese have ... explored to the fullest the uses and delights of ambiguity. One consequence of their way of handling the world is a tendency to leave unresolved apparent contradictions that would in other world views seem to cry out for resolution' (1974: 215).

If, in terms of time, the contradiction in the present context is still that signalled by Leach between *repetitive* and *non-repetitive* events, it is essential to conceive of it also in terms of other binary oppositions. That between Buddhism and Shinto proves to be particularly instructive. Buddhism is pre-eminently a religion of repetition, a tradition going back to its Hindu origins. There are four stages in life: conception, birth, death and rebirth. One can escape from this cycle, but if so, the fact is known seven days after death. If not, then rebirth occurs forty-nine days after death,[25] and according to Japanese belief the deceased becomes a *hotoke*, or buddha, a transformation given ritual recognition in the next following Bon festival. This is the Buddhist meaning of the domestic rites of *shonanuka* and *chūin*, which I have described above. In numerical terms, the Buddhist connection is firmly established by the number 7, which in Japan hardly ever occurs outside Buddhism (Smith 1974: 51). At the same time the end of the 49 day period is called *imiake*, 'the lifting of pollution', and the association between death and pollution is pure Shinto.[26] In principle Shinto is a religion of non-repetition, regarding history as a one-way process originating in what is seen as an historical act of creation.

The greater part of Japan's recorded history has been a story of the conflation of Shinto and Buddhism, in a process known as *Shinbutsu shūgō*.[27] Needless to say, this involved reconciling the non-repetitive time of Shinto, with the repetitive time of Buddhism. No one solution to this problem was found, and in the words of Smith cited above, every solution left 'unresolved apparent contradictions'. The solution I give below is based on two modern Japanese texts, Tsuboi (1970) and Miyako (1980), but it cannot claim to be anything more than one possible model, in a form generally acceptable in modern Japan.

If there is one governing principle, it is that which Caillet has called the 'interchange of temporal segments' (Caillet 1986: 32). This supports the conclusion I reached in my linguistic analysis, that in Japanese cognition, time is seen in terms of *alternation* rather than of any *cyclical* movement. The analytic framework is given in figure 7.2. The horizontal axis, defined by birth (*tanjō*), and the first rites after death (*sōshiki*) separates the world of the living from that of the dead.[28] The vertical axis, defined by marriage (*kekkon*) in this world, and separation (*tomurai-age*) in the other world, divides the combined state of life and death (*seishikan*) into two domains, one Shinto and the other Buddhist. The two axes combine to define four quadrants, which are alternatively stable

*Figure 7.2* The ritual cycle of life and death

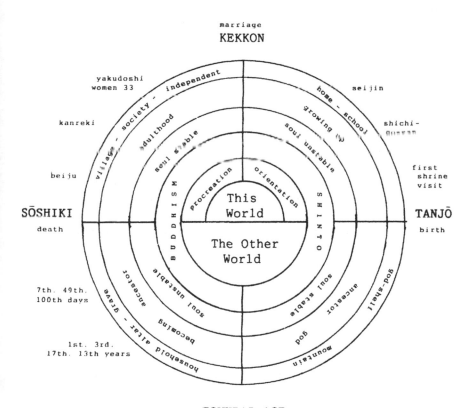

(*antei*) and unstable (*fuantei*). To the two unstable quadrants belong both the succession of Shinto rites following birth and that of the Buddhist rites following death. These rites, in terms of the days and years in which they take place, are defined numerically, in the manner already described. The periods in which they take place are essentially transitional, and may be seen as leading, ideally, to the stable periods which then follow.

Stability is best defined in terms of becoming a *soshin*, or 'ancestor god'. In this world, *kekkon*, or 'marriage', is the crucial event which makes this possible, for in marriage is the potential to succeed to, or found, a household (*ie*), whose future members will ultimately worship one as an *ancestor god*. In the other world, *tomurai-age* is the essential rite of separation, defining the end of the series of Buddhist mortuary rites, thirty-three or fifty years after the death. This is the signal for the memorial tablet (*ihai*) to be removed from the family ancestral altar, to mark the transition from buddha to god (Smith 1974: 96), or in the appropriate Japanese terminology, from *hotoke* to *kami*. Seen in these terms, just as *kekkon*, in *this* world, is the transition from Shinto to Buddhism, so also is *tomurai-age*, in the *other* world, the transition back from Buddhism to Shinto.

This final transition has two quite general aspects – loss of identity and return to nature – which manifest themselves in different ways throughout Japan. The removal of the *ihai* effaces the identity of the ancestor within the household, which may be achieved by the ritual effacement of the posthumous name. The return to nature may then be realised by throwing the *ihai* into a river, from which a pebble is taken to be placed in the shrine of the tutelary deity, but more often the *ihai* is deposited directly in the shrine (Smith 1974: 97). Either process is sufficient for the transition from Buddhism to Shinto, which in one explanation cited by Smith is described in the words *hotoke sama ga karada o aratte kami sama ni nari*, or, 'the buddha washes its body and becomes a god'[29] (Oshima 1959: 94). The ritual purification of the dead, which Shinto requires, is thus secured.

The problem here is that one is brought back to the non-repetitive time of Shinto. The ancestor god is *not* reborn. It is at best a *kami* who will return at *shōgatsu*, the new year's festival in honour of the ancestors. But it will be not only a *kami*, but also a buddha, or *hotoke*, who will be called back to its old household every year at the beginning of the Bon festival, now held from 13 to 15 August. The status of *hotoke* is achieved much earlier in the series of mortuary rites, for this is the purpose of the ritual (which I have already described) of *shōrō okuri* on the day after the end of Bon in the year following the death. The new, or *nii-hotoke*, is then incorporated into the collectivity of ancestors which the community, and not the household, call back to their homes for the period of Bon. This then means that the repetitive time of the community runs parallel to the non-repetitive time of the household. At the same time, however, the household also provides for its own welcome to its ancestors, so that Bon is one of the instances of a bridge between two different times, when the 'human and divine worlds come into contact'[30] (Caillet 1986: 35).

The annual agricultural cycle can also be interpreted in terms of binary opposition, defined in terms of the infinite succession of spring and autumn, separated by an axis with the Shinto new year festival of *toshigami* (literally 'year gods') at one end and the Buddhist Bon at the other. The former corresponds to the birth of the grain spirits (*kokurei tanjō*) and the latter to the harvest festival, (*shūkaku matsuri*) after which the grain spirits are at rest. This is illustrated in figure 7.3. Once again there is an alternation of stable and unstable periods,[31] with the latter, in the case of spring, being marked by a series of operations, such as sowing, transplanting, weeding and the extermination of pests, each with their ritual counterparts having a precise date in the lunar calendar.[32] In this case also, the series of events is defined *numerically* in relation to time.

Finally, Shinto contains a number of instances of a long-term cycle determining the parallel destruction and reconstruction of shrines. The best known case is the *naigū* at Ise, the shrine where the sacred mirror of the sun goddess, Ameterasu Omikami, is kept. This shrine alternates between an eastern and a western site, and in 1993, the present building on the western site will be replaced by an identical building on the eastern site. This symbolic act of renewal, completed in the month of October, occurs every twenty years, and in such a year determines its own ritual calendar.[33] The significance of the number 20 is not clear, since for other shrines the period is of a different length. The Nukisaki shrine in Gumma prefecture is rebuilt every thirteen years, the Kasuga shrine in Nara, every thirty years, and the Kamo Mioya shrine in Kyoto, every fifty years. What is significant in all these cases is that the constantly repeated rebuilding keeps alive an ancient tradition, but expresses it in a concrete form which is always new. At Ise, for instance, there is no conflict between the newness of the buildings and the antiquity of the shrine. The latter corresponds to that of the sacred mirror, which it always houses, so that, with every rebuilding, one of the most important rites is the transfer of the mirror to the new building.

*Figure 7.3* The production cycle in the perspective of life and death

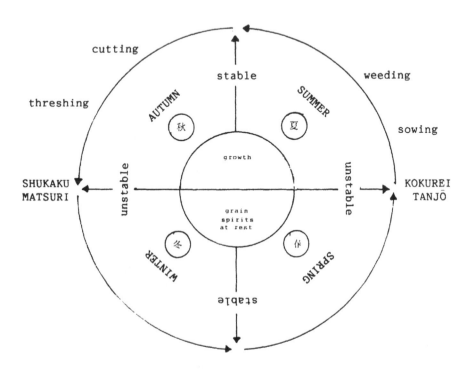

## THE CALENDAR: ITS USE AND MEANING

The numerical basis of the Japanese calendar, in any of the versions used in the course of history, has always been quite explicit. This point hardly needs to be restated. Only two quite straightforward questions need to be asked. The first is, what numbers are used, and the second, how are different units of time co-ordinated? As to the first question, the answer is simple enough when it comes to numbering days and months. Days are simply counted from the beginning of the month, months from the beginning of the year. The principle remains the same, whether it is a question of the lunar calendar, which was used officially until 1873, or the solar (Gregorian) calendar, which then replaced it. The principle itself is not affected by the fact that the first day of the lunar month was determined by an astronomical event, that is the new moon. The calendar, in any version, is no more than a system of counting, in which the numerical base is determined to accord with the phases of the moon and the earth's orbit round the sun. The numerical consequences are, however, significant, in that such a system does not lend itself easily to even the most elementary arithmetic. So long as the calendar retains an astronomical basis, the calculation of the exact length of time between any two dates can never be straightforward. This means in practice that the primary function of cardinal numbers, as I describe it in chapter 1, is suppressed. This explains why, historically, the calendar is an institution of ordinal numbers. Once this is accepted as a matter of principle, then the actual choice of the system of numeration is largely freed from the need to satisfy arithmetical criteria. The true merit of the system is then to locate, and order, events in the time scale with unique and systematic precision. This is certainly the position taken in Japan.

The question of the choice of a numerical system becomes much more problematical when it comes to the numbering of years. There are essentially two solutions to the problem, both of which are applied, in different contexts, in Japan. The first solution is linear, the second, cyclical. The Gregorian calendar applies the former to the numbering of years, and the latter to the numbering of months in a year and days in a month. The same is true of the official Japanese calendar, but at popular level the Japanese have also used the sixty-year cycle based on the Chinese *kanshi* matrix, introduced in chapter 3. Both systems require further discussion.

The first problem which any linear system must solve is to establish a base line. The Gregorian calendar, based on the nativity of Christ, simply counts the years of the new era from that starting point, just as the Islamic calendar[34] starts with the historical event in the life of Mahomet known as the *hegira*.[35] The alternative, adopted by the Japanese, is to go back to beginning of time, represented by what is claimed to be an *historical* act of creation.[36] This, for the Japanese, is taken to be the installation of the first emperor, Jimmu, in the year 660 BC.

Although a modern Japanese almanac will give the current year both as 1990 (*seiki*, or 'western annals') and as 2650 (*kōki*, or 'imperial annals'), neither version is true to either official practice, or the historical record. In both these cases, the counting of years starts again at the beginning of every succeeding era, a practice of undoubted Chinese origin.

The basis of the system is a succession of names known in Japanese simply as *nengō*, which only means 'year title'. In China the imperial court could decree the start of a new era to begin with any new calendar year, and this was frequently done both for political and astrological reasons. A rule that no name should ever be repeated ensured for every year a unique designation, in which the era name was combined with the number of the year in the era.

This system was adopted by Japan in the year 645, in which a coup d'état led to the enthronement of the emperor Kōtoku. The first era name *Taika*, meaning 'great reform', was then decreed to mark this significant political event. From the adoption of the *nengō*, *Taihō*, in 701, the fifth year in the reign of the emperor Mommu, the system was followed continuously until 1868, the year in which the emperor Meiji succeeded his father, Kōmei. In this period of more than a thousand years, the reign of eighty successive emperors was divided into 230 eras, each with their own, never to be repeated, name, so that the average era lasted scarcely five years. Even if there was no loss of either certainty or precision, this method of dating was hardly transparent, to say nothing of the demands which it made upon memory.[37]

One of the reforms following the Meiji Restoration was that the era name chosen by a new emperor should remain unchanged throughout his entire reign,[38] and then, after his death, become his posthumous name by which he would be known to history.[39] This explains why the late emperor, known in the west as Hirohito,[40] is now simply *Shōwa Tennō*, that is, the *tennō* of the *Shōwa* era (1926–89). It was, however, only in 1979 that the Diet passed legislation making the use of the *nengō* in its revised, modern form, mandatory for official purposes.

The Japanese era names have always consisted of the *on* readings of two *kanji*, which must be selected from a list of seventy-two recognised characters. The selection cannot be arbitrary, for the combination must have some meaning according to one or more of a small number of recognised Japanese or Chinese classics.[41] In the case of the present emperor, the name chosen, Heisei, was one of three possibilities.[42]

To summarise, the naming of years in Japan follows three different systems. The first depends on the names of the emperors before the time that the *nengō* system was established in the seventh century. The second is then the *nengō* system as it was adopted from China, and the third, the modified form of this system adopted from the beginning of the reign of the emperor Meiji in 1868. If the third system is in principle a return to the first, there is an important practical difference, in that the record of the time of the first system is not strictly historical. The sixth emperor, Kōan, according to this system, reigned from 392–291 BC, a period of ninety-nine years, and this is but one of the periods which historians must now reject.[43]

Neither of the three systems was ever intended to provide the basis for calculating long periods of time, and before the Meiji era this was never considered important. When, however, Japan entered the modern world, its governors set great store by the antiquity of the imperial line, and by a process of elementary, if laborious arithmetic, they reckoned that the imperial line had already lasted for more than two and a half millennia – not that this period had any numerological significance in the Japanese tradition. By this reckoning the imperial line was due to celebrate its 2,600th anniversary in 1940. The exact date would be 11 February, the public holiday known as *kigensetsu*.[44] The celebrations, presided over by the Shōwa Emperor, lasted two whole days,[45] for which the appropriate ritual, comprising both Shinto and popular elements, was devised.[46]

The sixty-year cycle based on the Chinese *kanshi* matrix is both more transparent, and more evocative,[47] than the official system of counting years in every successive *nengō*. This explains the extraordinary popularity it has always enjoyed in Japan as an alternative to the official system. If, today, it is largely important for its use in horoscopes (which I describe in chapter 6), it still has a number of quite practical advantages over the official system. Of these the most important is the precision with which a year can be referred to in any sixty-year period, which in many contexts need not be made explicit. If I were to tell a Japanese acquaintance that I first visited his country in a *kōshin* year, and that I was planning my next visit for a *kanjutsu* year, he would have little difficulty in knowing that the former year was 1980 (and not 1920, 1860 and so on into the indefinite past), and the latter, 1994 (and not 2054,[48] let alone 2114). And even then, if I had wished to be quite explicit about the year of my first visit, I could always have said 'Shōwa kōshin'[49] – the traditional Chinese form.[50] In the case of future years, also, the installation of a new emperor does not matter, whereas the official system carries the necessary implication that the present emperor will live into the indefinite future.[51]

The *kanshi* cycle can also be used for counting months, days, and even hours. In the case of months, the twelve *shi* will simply maintain the same cycle in every year, and the same is true for the twelve hours of the day.[52] Days are also counted according to the *kanshi* cycle, but in this case every year has its own allocation, which will repeat itself only at long and irregular intervals. The combination of the *kanshi* representations for the year, month, day and hour will always produce an eight-character code, which was used for the purposes of divination according to the ancient Chinese almanac of T'ung Shu (Palmer 1986: 38). In practice the modern Japanese almanacs are only interested in the *kanshi* cycle of days. These play a part in horoscopes, together with the cycle based on the twenty-eight constellations of traditional Chinese astronomy. The latter in turn divide into four 7-day cycles, in which each day is associated with the sun, the moon, or one of the five planets known in the ancient world.[53] This is the basis of the week, as it is known in Japan , but although the seven-day week now rules in modern Japan as much as it does in the west, it still remains an entirely profane institution. The question, already asked, about its significance in Japanese life as the basis of a cycle whose length always remains fixed, must be answered in purely pragmatic terms. The week belongs to that part of Japanese life which is governed by the social, economic and technological institutions derived from the western world.

This, also, is the world of that universal Japanese timepiece, the quartz watch, which measures time with unprecedented accuracy.[54] Time, measured in these terms, determines the punctuality of the trains, and other forms of public transport, which the Japanese count upon to run *on time*. This is the apotheosis of *arithmetical* time, which (with its use of cardinal numbers) makes possible such claims as that the bullet trains of the *shinkansen* run from A to B, a distance of N kilometers, in T minutes, at a breath-taking average speed of S kilometers per hour, where one may be assured that $S > 200$.[55] One is left to ask whether this is the ideal of the new Japan. Perhaps so, but then Ohnuki-Tierney (1984: 175) cites the general lack of any appointment system for medical treatment as reflecting the Japanese concept of time.

What then is this concept? Why is the calendar acceptable, and not the clock? Caillet (1986) points the way to an answer. First, 'the Japanese have continuously shown a marked lack of interest in the *computation*[56] of time and, generally speaking in astronomy: they seem preoccupied only by the astrological use of calendars' (ibid.: 31). This immediately establishes a sort of moral superiority for the lapse of time over the year, as against the day. The Japanese give priority to the 'natural and social realities, the harmony of which the calendar expresses. Yet, this harmony does not seem to be realized by a regular *mathematical* symmetry, but rather between the constant relations between the human and divine worlds' (ibid.: 44). It is these relations which link the *repetitive* time of the calendar, to the *non-repetitive* time of the life of the individual. The point is that, according to the Japanese concept, it is not required that 'a central subject compose the unity of time; it is, on the contrary, quite satisfied with the extreme mobility of the subject on the *surface* of time' (ibid.: 45).

The calendar, seen from this perspective, is a record of what the individual observes, not only from nature and the cosmos, but also from contexts in which the gods become manifest. That such contexts can be functional, particularly in their relation to the annual agricultural cycle, when they may determine the timing of a whole succession of operations, sowing, transplanting, irrigation, harvesting and so on, implies no essential contradiction. In a religion such as Shinto, which is so closely tied to the relation between man and nature, one could hardly expect otherwise. Seen in this way, the mathematics of the calendar is a sort of epiphenomenon, which, however indispensable for its use in practice, does not represent its true character.

If this view is correct, what then of the use of calendar for horoscopes, which, as I have shown in chapter 6, is almost entirely arithmetical? The answer must be found by means of a sociological analysis. Disdain for arithmetic is characteristic of the elitist samurai tradition of unreformed Japan as a rural and feudal society (Nakayama 1969: 161). In this context, only the mathematics of astronomy was worthy of any respect (ibid.: 160). For the rest, mathematics was considered 'a mercenary activity, fit only for traders and petty officials' (ibid.: 159), who were precisely the class of people who used the abacus.[57]

The whole history of modern Japan must be that the mercenaries and petty officials came to dominate society, while the samurai have passed into oblivion. The battle was hard fought, and the old ethic still survives. In the use and interpretation of the calendar the conflict *still continues*.

# 8 The spatial world of numbers

## IN ONE DIMENSION

Numbers, in the dimension of space, can be used in two ways. The first is to name, and implicitly to order, recognised points, the second is to measure the distance between them. The first use is of ordinal, the second, of cardinal numbers. These alternative uses suggest an analogy with time, where points, *jiten*, are separated by periods, *jidai*. The analogy is, however, imperfect. It implies that the properties of space, or *sora*, are comparable to those of time, or *toki*. Noting the *on* reading, *kū*, corresponding to the *kun* reading, *sora*, one would expect some conceptualisation of space in terms of *kūten* and *kūdai*, but in fact these words do not exist. Nor are there any other Japanese words which convey the meaning sought after. True there is *chiten*, literally a 'land-point', but this depends upon space reduced to two dimensions. This is the nub of the whole problem. Space, as such, is too inchoate to support any sort of order, whether or not it is numerical.[1] The ordered movement of the celestial bodies, which is critical for the measurement and definition of time, is a meagre basis for organising space at terrestial level.

The problem has both a static and a dynamic aspect. The former is apparent in the domain of architecture, in which numerical considerations govern the design and location of buildings. The problems which arise are two- or three- dimensional. The dynamic aspect, defined by *movement*, is essentially one-dimensional, and this being the simpler case, it will be considered first.

The numerical organisation of one-dimensional space reduces, at an elementary level, to assigning numbers to points on a line. The question is then, which line? In contrast to time, experience of nature or the cosmos indicates no single line on which such points must occur, nor any means for identifying their location. The choice, in both aspects, is left to man – in other words, it is a matter for the local culture, to be studied in relation to human geography. Nor is it a question of one single line: there can be any number that are culturally significant, and this is certainly true of Japan. Viewed topographically, the lines occur in a two-dimensional space, but this only becomes significant when points of intersection have to be dealt with. In the history of almost any culture this is rather a refined case, typical of the interchanges in a modern transport system, but much less important in the traditional culture.[2]

From this perspective there are essentially only two cases: the line between two points, and the line which closes a circle.[3] In the latter case there may be the additional problem of choosing a point on the circle as a starting point. In both cases it is movement along, or around the line, which makes it *culturally* significant. There is therefore a *why* as well as a *which* question to be answered.

Historically the best-known Japanese case of a line joining two points is the famous Tōkaidō, or 'East–west road', originally joining Tokyo to Kyoto, but later extended to Osaka. Numerically this is interesting for the fifty-three *tsugi,* or 'post-station towns', which offered services to travellers. The woodblock prints of Hiroshige, and other works of art and literature, made these so well known as to define the meaning of 'fifty-three' in Japanese culture. There is no evidence that the number was deliberately chosen. The road followed a natural route along, or close to the Pacific coast, and the *tsugi* were merely located at convenient intervals.

The road is a particular case of a line between two points. More generally, such a line is equivalent to the Japanese *sen* (線), such as occurs in the familiar *shinkansen* (literally 'new trunk line') used for high-speed trains running between the larger cities. The two cases (introduced in the last paragraph but one), of a line between two points and a closed circuit, can be modelled on two actual instances: the first is the modern Kyoto *chikatetsu,* or 'underground railway', in its original form consisting of a single line with eight stations; the second is the much older Osaka *kanjōsen,* or 'circle line', with seventeen stations.[4]

The numerical ordering of the systems is for those who travel by them far more important than their actual length. Who is Kyoto knows that the *chikatetsu* is 6.6 kilometers long, or in Osaka, that a complete circuit of the *kanjōsen* is 21.7 kilometers?[5] Much more useful is the knowledge that the former operates on a shuttle lasting fifteen minutes in each direction, and the latter of a circuit repeated every forty minutes. The disordered realm of space has been reduced

to the ordered realm of time. The points in space are not even
numbered according to the ordinary Japanese numerals, although the
names of the successive stations can be seen as a system of *meta-
numbers* such as I describe in chapter 3.[6] In the case of the Osaka
*kanjōsen* this has the advantage of avoiding the necessity to identify
one particular station as the starting point.[7]

The conquest of space seen in terms of time is a characteristic of
the restless modern age, when travellers in their millions must cover
so many points along a *sen* simply to perform the duties of their daily
life. In the urban culture of modern Japan, travel has become
profane, and its ideal component is speed, the factor which reduces
space to time.[8] It is not for nothing that the only instrument on the
wall of the buffet-car of the *shinkansen* is a speedometer, which as it
measures an almost constant speed of more than 200 km.p.h. re-
assures the impatient *sararīman* that he will not be late for his
appointment. In the same mode, the departure time of the train
becomes effectively its name, once more an instance of numbers
being used for metonymy.[9]

In the rural culture of traditional Japan travel outside the domain
of the village was not part of daily life. Those engaged in non-
agrarian occupations which required travel from one place to another
for the performance of the services offered have long been regarded
as having special status, whether they be religious ascetics engaged in
continuous pilgrimage (Blacker 1986: 167), craftsmen or simply
entertainers.[10] The world they inhabit is *soto*, or 'outside', in contrast
to the inside, or *uchi*,[11] realm of the ordinary villager (Berque
1987: 12). This is also the world of the *yama*, or mountains,[12] where
the *kami* dwell, so that on certain prescribed days of the year the
people may proceed in pilgrimage to the top of the sacred mountain
above their village. There they will find a shrine, where the appropriate Shinto rites will be carried out.

The pilgrimages may go much further afield to such sacred mountains as Fuji, Haguro or Ontake. The routes to the summit of the holy
mountains are divided into ten stages, separated by points known as
*gōme*, where the climber will find a hut for rest and refreshment. For
a small fee the guardian will also brand the pilgrim's staff with the
symbol for the stage completed, so that in the end the climber who,
completing the tenth stage, reaches the top of the mountain, will have
a record of his achievement.[13]

Covering the ten stages is likely to be a walk up-hill helped by
occasional flights of steps when the route would otherwise be too
steep. This is a reification of the principle implicit in the word '*dan*',

with all that it means for achieving merit by means of ascent to progressively higher levels. *Dan* imports numerical ordering. The advantage of a mountain pilgrimage is that anyone, with sufficient energy, can reach the highest level in a fixed number of steps, and in Japan, at least, very few give up on the way.

At the same time the summit represents the centre, or the core (*la*) of the *mandala*. The mountain itself is taken as a *natural* symbol of the Holy Mount Sumeru, which is the ideal image of the *mandala*. Although there may only be one recognised route to the summit,[14] this does not detract from the image. In the case of Fuji the centre, geographically, is the core of the extinct volcano. Climbers often circle the entire rim – a distance of a mile or two – as if this had some ritual significance,[15] but there are no stations marking out the circuit.

In addition to mountain pilgrimages, which require going to and returning from a fixed point – the top of the mountain – pilgrimages which require completing a circuit of a fixed number of points are equally well known in Japan.[16] In such a case the pilgrimage can be joined at any point, and although it is customary then to proceed in a clockwise direction, this is not obligatory (Reader 1988: 52).

Two such pilgrimages are particularly well known. The *sanjūsan-kasho*, starts at the temple of Seiganto on Mount Nachi in Wakayama prefecture, to make a broad circuit round Kyoto, including a number of temples in and around the city itself. The name means 'thirty-three places', and this is the number of temples, all dedicated to the Kannon Buddha, to be visited. It was believed that the pilgrim who visited them all would be safe from hell. The number 33 is chosen for the same reason as it is in the Kyoto temple of Sanjūsangendō, which I consider in the following section. This, then, is the basis of the so-called Saikoku, or 'west country', pilgrimages. The *hachijūhakkasho* works on much the same principle, with visits to Shingon Buddhist temples in 'eighty-eight places' along a circuit around the island of Shikoku.[17] The idea is that Kōbō Daishi,[18] the founder of the Shingon sect in the eighth century, accompanies every pilgrim (Reader 1988: 53).[19] In this case the reason for the number 88 is unknown.[20]

The interesting thing about the Shikoku (and to a lesser extent the Saikoku) pilgrimage is that it provides a model basis for other pilgrimages, in which the focus and symbolism, as well as the number of stages, are the same (Reader 1988: 58). As far as the mathematics is concerned it is the topology, based on the number 88, which is preserved, while the scale is much reduced,[21] although in one or two cases an attempt is made to maintain relative distances.

As an ascetic exercise a short walk around a small-scale pilgrimage route, sometimes capable of being completed within an hour, is hardly comparable to a complete circuit of the *hachijūhakkasho*,[22] but then the modern pilgrim will most likely travel this as a member of a guided bus-tour.[23] The apotheosis of this process is to be found in the numerous guided tours organised by railways and bus companies to shrines and temples in the locality dedicated to one or other of the *shichi-fuku-jin*, known in English as 'the seven gods of good fortune'. As noted by Reader, this is a heterogeneous 'collection of Japanese, Indian and Chinese deities with roots in Shinto, Buddhism and Taoism who have merged with Japanese folk belief' (Reader 1987a: 6). In this case the pilgrimages, if such they may be called, have no ascetic background whatever. It is not for nothing that Reader's article appeared in *Kansai Time Out*. In modern Japan, transformed into a fully literate urban society, religion becomes recreation.

The transformation is worth looking at in greater detail, particularly for its implications for the role of numbers in Japanese popular culture. The traditional position, in pre-modern Japan, is aptly described by Turner:

> As the pilgrim moves away from his structural involvements at home his route becomes increasingly sacralized at one level and increasingly secularized at another. He meets with more shrines and sacred objects as he advances, but he also ... has to pay attention to the need to survive and often to earn money for transportation, and he comes across markets and fairs, especially at the end of his quest, where the shrine is flanked by the bazaar and the fun fair. (Turner 1974: 182)

The world of money in traditional Japan was that of the lowest rank, that of merchants, in the social hierarchy. This was the world of the abacus, which enabled numerical calculations to be carried out to serve the practical demands of commerce. At this stage of cultural evolution, Turner sees pilgrimage as essentially liminal, 'the ordered anti-structure of patrimonial feudal systems. It is infused with voluntariness though by no means independent of structural obligatoriness' (ibid.: 182), and is 'an amplified symbol of the dilemma of choice versus obligation in the midst of a social order where status prevails' (ibid.: 177).

All this occurs in a world in which

> Daily, relatively sedentary life in village, town, city, and fields is lived at one pole; the rare bout of nomadism that is the pilgrimage

journey over many roads and hills constitutes the other pole ... the optimal conditions for flourishing pilgrimage systems of this type are societies based mainly on agriculture, but with a fairly advanced degree of craft labor, with patrimonial or feudal political regimes, with a well-marked urban–rural division but with, at the most, only a limited development of modern industry. (ibid.: 171)

That puts pre-modern Japan in a nutshell.[24] For contemporary Japanese, it is another world. Modern city life is anything but sedentary, and nomadism has become almost obsessional. Massive political support is necessary to ensure the survival of agriculture in a society with the world's technologically most advanced industry. Yet in a world where number-crunching computers operate twenty-four hours a day, it is still important not to miss out on a single one of the eighty-eight temples of the *hachijūhakkasho*. Indeed, modern technology assists the process with instant cameras which record the date of every picture,[25] so that the principle of the official seal, or *hōin*, is becoming almost otiose for the purposes of the record.

The explanation for all this is to be found in a sort of cultural regression: participating in an instant pilgrimage puts numbers back in their proper place. The individual who has completed the *hachijū-hakkasho* has established, to use Turner's word, *communitas* with his fellow travellers, because they share the number 88 in common. The pilgrimage has somehow reinstated the symbolic power of the number, which, for the Japanese participants is almost beyond value.

## SANJŪSANGENDŌ

The well-known Kyoto temple of Sanjūsangendō has already been introduced in chapter 1 in relation to the definition of the *nombre marginal*. At the same time, the temple, dedicated to the Kannon Buddha, would seem to mark a transition from the use of numbers in one dimension. None the less, it is still first and foremost a structure designed to convey a message based on numbers. In relation to the way this is achieved, the fact that the actual temple is in three dimensions, and is built on a two-dimensional ground-plan, is incidental.[26] The design both of the building itself, and of its contents, was governed by a concatenation of numbers occurring in the Lotus Sutra.

The basic concept of the building is one dimensional. It consists of a pavilion, built on a north–south axis, open on the east side, and containing a row of thirty-three bays which give it its name, *Sanjū-*

*san*gendō. The building contains 1,001 statues of the Kannon Buddha.[27] These are identical, except that one of them, which occupies the central position in the centre bay, is much larger in scale. The remaining thousand statues are placed in ten rows of a hundred, so that they cannot exactly be divided among the thirty-three bays. The open plan of the building does not require this. Each statue has forty-two hands, and eleven small heads surrounding the main head. The numbers, then, that are represented by the building and its contents are 11, 33, 42 and 1,001.

The number 33 is derived, in first instance, from chapter 25 of the Lotus Sutra, which is known to the Japanese as the Fumombon.[28] This chapter, after describing a number of catastrophes in which the Kannon Buddha, known also as the 'Regarder of the Cries of the World',[29] will come to the rescue, lists thirty-three bodies in which the living will be saved. It is also the number of the different figures into which the Buddha can be transformed (with the result that the 1,001 statues represent 33,033 separate manifestations). Being a number one greater than a power of 2, 33 is also that of a simple *mandala*, as defined in chapter 3. As such it represents the heavens at the top of the mystical Mount Sumeru. Here, one heaven, that of Indra, who rules the mountain, is located on its central peak, leaving the remaining thirty-two to be divided among the cardinal points following the symmetry of the *mandala* (*The Threefold Lotus Sutra* 1975: 147). As a *nombre marginal,* 33 is derived from the 32 signs distinguishing the body of the Buddha (ibid.: 7f), for by adding 1 to this number, with its earthly connotations, one enters into the realm of *unbounded* perfection.

Where 33 has to be expressed as 100001 in the binary system to make clear its status as a *nombre marginal,* this status for 1,001, which counts the actual statues of the Kannon Buddha in the temple, is clear on the face of the number itself. To find its significance, one must look, in this case also, at the number which it exceeds by 1. This, 1,000, counts the smaller statues. Its numerical sigificance comes from the fact that $1,000 = 25 \times 40$. The first number in this product simply represents the twenty-five sorts of the life and death of man. The meaning of the second number, 40, is more involved. In principle each statue should have a thousand pairs of hands, but since each actual hand is taken to represent twenty-five, only forty pairs are necessary. For every statue this number is increased to forty-two, by two additional pairs of hands which are not counted because they have a special function. Of these one pair is clasped in prayer, while the other holds a small bowl.[30]

The number 11, of the smaller heads surrounding the head of Buddha in each statue, has as its arithmetical basis, the fact that $11 = 3+3+3+1+1$. Symbolically this represents three heads looking forward to mean compassion, three looking to the right to mean understanding, three looking to the left to mean anger, one looking backwards to mean laughter, and finally one looking upwards from the top of the head to mean perfect understanding. Although the pattern lacks the perfection of a *mandala*, the heads are seen to face in all directions. This shows that the Buddha, embracing all moral attributes, is responsive to all the cries of the world, which accords with the description in chapter 25 of the Lotus Sutra.

If 11, 33, 42 and 1,001 are the numbers expressly represented by Sanjūsangendō, it is worth asking if the fact that 3, 7 and 11 constantly occur as factors is significant. As noted in chapter 7, the number 7 has particular associations with Buddhism, but this is no proof that it is particularly significant in relation to Sanjūsangendō. The significance of 3 and 11 in this context is even more uncertain: in any case the four base numbers, 11, 33, 42 and 1,001 are sufficient for the essential message conveyed by this remarkable building, and its even more remarkable contents.

## TWO AND THREE DIMENSIONS

Historically, the use of numbers in the ordering of space in two or three dimensions is governed by the distinctive ecology of Japan. The country, in its natural state, is mountainous, with a dense forest cover. Wood therefore is the obvious construction material. At the same time, rich volcanic soil and a warm temperate climate with reliable rainfall make the country well-suited for wet rice cultivation. The material implications of these factors still govern the use of numbers in two or three dimensions.

The terraced cultivation of rice is to be seen almost everywhere in Japan in the familiar *dandanbatake* on the hillside above the villages. This is numerically significant for dividing the continuum into a finite number of discrete intervals. It would be convenient to see this as the model basis for the steps cut into the side of sacred mountains, but then there is the problem that the characteristic fan layout of the terraces is not linear. In practice the terraces belonging to any one village were named, but not numbered, and the names were sufficient to identify them with their respective owners. If the geography of the terraces seldom lent itself to systematic numeration, the same was true also of the distribution of the houses in the village. Although, in deciding upon the orientation and location of a house, the principles of geomancy could not be neglected, the natural constraints of the terrain often left little room for choice.

The form of the house itself was largely determined by the fact that it was built of wood. The essential constraint imposed by wood is that the ground-plan must be based upon a rectangular grid.[31] This has been characteristic of the standard Japanese family house, or *minka*, from the time of the very earliest records. The traditional Japanese building always had but one main floor (Masai 1987: 67), and the floor-plan was based either on an interpost-span module or a *tatami* module (Itoh 1972: 111). In the simplest case, such as that of the fourteenth-century Rin'ami house (of which the floor-plan is now kept in the archives of the Toji temple in Kyoto), the house was a simple rectangle, so that the Rin'ami house had four bays along one side and six along the other (ibid.: 138).

The familiar *tatami* mat was already at this early stage the key to the ground-plan.[32] Although the size of the *tatami* has never been entirely standardised, they are always about 6 feet long, and 3 feet wide. Because in every case the length is exactly twice the width, they provide the modular basis for any room whose dimensions are based on fixed multiples of the length of the shorter side,[33] as illustrated in figure 8.1. This is now standard in Japanese domestic architecture, although there are still small variations in the size of *tatami* between different regions.

The modular principle established by *tatami* is essentially two dimensional. The construction of the traditional Japanese building does not generally allow its application to be extended into three dimensions, but there are occasionally instances of this which occurred, historically, by force of circumstance. One of the best instances is provided by the platform supporting the main buildings of the well-known temple complex of Kiyomizu in Kyoto, which are located on the side of a steep hill. The support for this platform consists of a three-dimensional complex of posts and beams, whose design is clear to any visitor. This is in the form of a three-dimensional modular grid, in which every module has the same length, width and height. The number of modules in each dimension seems to have been determined purely by engineering principles, and no special numerical significance is claimed. Indeed, many of the beams and posts are greatly foreshortened to take into account the profile of the hill on which the temple is located (and which explains the necessity for the whole structure in the first place).

In Japan the rectangular modular plan for buildings has seldom

*Figure 8.1* Japanese floor-plans based on interpost-span module (top) and
*tatami* model (bottom)

*Notes*:  a = 1) length of one *ken*, 2) distance from centre of one post to another, 3) standard unit of
measurement. *Tatami* for any given room are all of the same size, but the size may vary from
room to room.
b = 1) length of one *tatami*, 2) standard unit of measurement. Size of *tatami* is uniform
throughout the plan.
卐 = Buddhist altar.

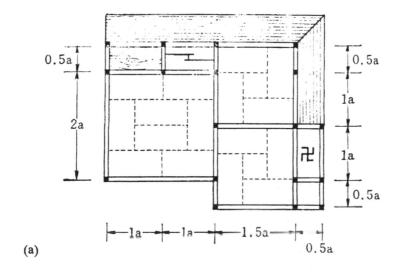

(a)

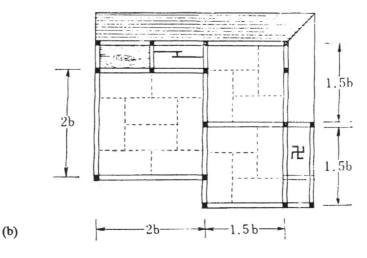

(b)

been used for planning on the scale of a whole town. It is standard for any number of shrine and temple complexes, although in such cases it may arise from the necessity to achieve the correct north–south orientation of the buildings.[34] The use of such a plan for cities is far from standard, but there are two exceptional cases, one modern and the other ancient.

The former is Sapporo, the capital of the prefecture of Hokkaido, which embraces the whole of the most northern of the four main Japanese islands. The colonisation and development of Hokkaido is part of the history of the present century, which explains the circumstances in which it was possible to lay out the capital city on a rectangular grid. The model could hardly be simpler. The streets in both directions are numbered according to their distance from two central axes. The model is that of a graph, with x- and y-axes, and Cartesian co-ordinates. The point is significant for the Japanese numerical tradition, since the application of the basic ideas and notation of algebra (conceived of as a general, abstract model of *numerical* equivalences) to geometry, which by definition is based on configurations in two or more dimensions, was only achieved by Descartes at the beginning of the seventeenth century (Williams 1978: 16). This type of *analytical* geometry only became known in Japan when the country became fully acquainted with western science in the later nineteenth century.

Traditional Japanese culture was if anything hostile to any such mechanical approach to town planning, and the results of this attitude are still to be seen in the chaotic street plans of almost any Japanese city, starting with Tokyo. Any sentimental attachment to a 'home town', however large or small, was based on the concept of *kokoro no furusato*, literally 'the old country of one's heart', or by implication the village of one's ancestors. From this perspective form (*katachi*) is less important than sentiment (*kokoro*), so that the spiritual domain is seen as hidden, profound, and essentially *mukei*, that is *without* form, where the material domain is *ȳkei*, or *has* form. According to this way of thinking, the contrast to be made in looking at Tokyo is between the slopes of the Yama no te literally 'hand of the mountain', and the low-lying Shitamachi, but such a view imposes no numerical order on town-planning. Only in modern times is there some semblance of order in the forms of addresses, so that the whole of Japan is divided into nearly a thousand postal districts, each with its own three-figure code, whereas at local level a specific building will be identified with a two- or three-part numerical code, such as 2-182, for a house in the Kyoto *chō* of Fūjiyama.[35] This process is taken to it utmost extreme with telephone numbers, where the model of the system represented by the circuit diagrams used by the engineers is no more than a remote abstract from the actual topography.

The Japanese distaste for Cartesian systems of town-planning had no effect on the layout of the ancient capital city of Kyoto. This is the second of the two exceptional cases already mentioned. The town-plan of Kyoto is a near perfect rectangular grid, and the streets running east–west are numbered from south to north, using the suffix *-jō* (条) as a counter.[36] The model was borrowed from China, as was also the rule according to which the ancient capital was divided into wards. The basis is the *mandala* so that the centre, Nakagyō, is surrounded by the *ku*[37] of Kita (N), Higashi(E)yama,[38] Minami (S) and Nishi (W). Looking at the city from the north provides the orientation for the remaining four wards of Ukyō (*right*), Sakyō (*left*), Shimogyō (*lower*) and Kamigyō (*upper*).[39] The first two names look at the city from the traditional perspective of the emperor, who sees one of the two *ku* immediately below him on his right hand, and the other on his left.[40] The second two names, meaning 'upper' and 'lower' may relate back, through Chinese Buddhism, to the *egg of Brahma* in traditional Hinduism, with the visible, light, *upper* half being separated from the invisible, dark, *lower* half, by a flat plane representing the earth (Crump 1990: 141). In the Buddhist tradition, the visible, top half came to represent Mount Sumeru, which once again brings us back to the *mandala*, which is a binary but not a Cartesian model.

The conclusion must be that the use of numbers to *organise* space on a large scale is marginal in the Japanese use and understanding of numbers. Kyoto as a type of town-planning may be a model useful for general application (as in Sapporo), but its numerical base is rooted in a tradition which has been largely superseded. The Cartesian idea is reflected in certain modern situations, such as the cadaster of rice-fields in the extensive flood plain in Niigata prefecture, or the complex of oyster beds to be seen off the coast of the Kii peninsular, but instances such as these are purely practical, and hardly part of the numbers game.

# 9 The Japanese abacus

## DESIGN AND USE

The basic abacus consists of a rectangular frame, generally made out of wood, in which the two longer sides are joined by a series of equally spaced columns, on each of which is threaded a number – always the same – of beads. The actual operation of the abacus depends upon these being moved up and down by the user, so it is not surprising that the Japanese word *shuzan* means, simply, 'bead calculation'. The abacus, in its most elementary form, seems to be hardly more than a children's toy of a kind to be found in many different parts of the world. An instrument operating on the same principle could equally consist of pebbles to be moved up and down the columns of a matrix imprinted in sand, or along parallel grooves cut into a wooden board. Such isomorphic instances are well documented.[1]

The practical use of the abacus depends upon the convention that the beads in every column can represent any number from 0 to N, where N is the basis of the numerical system in use.[2] The decimal system, as it developed in China, required that N = 10, so that in principle the Chinese abacus should have ten beads in every column. In practice a horizontal bar divided the columns of the Chinese abacus into two parts, with the upper part containing two beads, and the lower part, five. The two beads of the upper part each represented five units, and a number was represented by moving the appropriate number of beads from the outside frame to the central bar, so that, for example, the number 7 was represented by one bead (= 5) above the bar, and two below.

On this principle of representation, a single column could, theoretically, represent any number from 0 (with all beads returned to the outside frame) to 15 (with all beads brought to the centre bar). The usefulness of the abacus depended on the convention that the columns not only had a unit value, determined by the position of the beads, but a place value in terms of successive powers to ten, with the highest power being that in the left-most column – just as it is in the place-value system of Arabic numerals.[3] This allows any Chinese numeral to be entered by representing the numbers of the *cyclic* system, as defined on page 26, on the *frame* of the abacus. That is, the succession of columns defines the *frame*, while the beads on each separate column represent the *cycle* of numbers from 0 to 9. The actual definition of the powers to ten represented is left to the user, and depends upon the calculation being carried out. The same configuration of beads could represent, for example, 75,300 or 7.53, or indeed any number of the form $753 \times 10^n$, with $n$ being any integer, positive or negative.[4] The only basic rule is that the frame, once defined by the user, cannot be changed in the course of the calculation being carried out. This requires some care, since the actual definition of the frame, in any given case, is purely conceptual,[5] as opposed to the visual representation of the numbers of the cyclic system by the beads.

The form of the abacus restricts its basic use to the arithmetical operations of addition and subtraction, which are subsumed in the Japanese compound, *kugen*. This restriction follows from the fact that the only physical operation possible is to move a bead either towards, or away from the bar, the former operation being addition, and the latter, subtraction. The only additional rule is that every time *ten* beads occur on a column, they must be transferred to the column next to the left, to be represented there by *one* bead: this is no more than the essential operative principle of the place-value system. This, the rule for addition, has a complementary rule for subtraction that every loss of ten beads be processed by subtracting one bead from the next column to the left. On the Chinese abacus, which allows for any number from 0 to 15 to be represented by the beads of a single column, the process of conversion and transfer is palpable in the sense that 10, actually represented by $2 \times 5$ beads *above* the bar, can be subtracted by moving these two beads back to the frame, at the same time as 10, represented by a single bead *below* the bar, is added by moving this bead up to the bar, in the column next to the left. This establishes the principle that the operation of the abacus continually depends upon adding and subtracting the base number 10 in adjacent columns. All this is more or less isomorphic to the way in which addition and subtraction are carried out with Arabic numerals, when numbers are 'carried' from one column to the next.

In principle, if a number can be represented in either of two adjacent columns (such as the Chinese abacus allows for any number from 10 to 15), one representation or the other must be redundant. The present form of the Japanese abacus, introduced in 1935 (Kubo 1986: 184) (and illustrated in figure 9.1), puts this principle into practice, by means of the simple expedient of reducing the number of beads above the bar to *one*, and the number below to *four*. This means that any number from 0 to 9 can be represented on any given column, which is sufficient for any place value system based on 10, as the use of Arabic numerals demonstrates. (In the setting illustrated in figure 9.2, two numbers, 973 and 428 are represented.)[6] Once again, this is a case in which the system puts more demands upon the user at an elementary level, but at the same time gains in simplicity and speed of operation once the basic skills have been learnt.

In actual use, the frame of the abacus is held at the left hand end between the thumb and forefinger of the left hand of the operator. The thumb and forefinger of the right hand are then used for moving the beads,[7] and the user's first action is to grasp, with these two fingers, the centre bar at its left hand end, and then to move them, lightly, along the bar, to the right hand end. This simple action, completed in half a second, returns all the beads to the outside frame, so that the abacus itself registers zero. It is then ready to be used for a new calculation.

*Figure 9.1*  The Japanese abacus

*Note*:  The orientation points (*teiiten*) are small black points marked on the crossbar above every third column (*keta*), intended to help the user in laying out the calculation, but otherwise having no essential function. The *kanji* immediately above the frame is *ten* (heaven), and that immediately below, *chi* (earth); those immediately to the left and right of the frame are *ue* (upper) and *shita* (lower) respectively, referring to the descending powers of 10 represented by the columns reading from left to right.

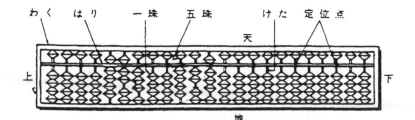

*Figure 9.2* Abacus multiplication
*Note*: The numbers in bold face are those that have been changed as the result of the operation.

| Initial stage representing the original sum | 9 | 7 | 3 | – | . | 4 | 2 | 8 | – | – | – | – | – |
|---|---|---|---|---|---|---|---|---|---|---|---|---|---|
| First step | 9 | 7 | 3 | – | – | 4 | 2 | – | **7** | **7** | **8** | **4** | – |
| Second step | 9 | 7 | 3 | – | – | 4 | – | **2** | 7 | **2** | **4** | 4 | – |
| Third step and result | 9 | 7 | 3 | – | – | – | **4** | **1** | **6** | 4 | 4 | 4 | – |

The user's next step is to decide how to allocate the columns of the abacus in the way best suited for the arithmetical operations to be carried out. In deciding on the best plan, the user is confronted by two restrictions inherent in the nature of the instrument itself. The first is that the abacus has no built-in store or memory. In principle every time a number is used in a calculation, it effaces not only itself, but also the result of all preceding operations. There is no built in way of recording any interim result.[8] The user can overcome this limitation by allocating a number of columns as a storage facility, and some calculations require this. This procedure, however, increases the effect of the second restriction, which is simply the limit to the number of columns on any given abacus. True this can be overcome by simply using a larger, or better, longer abacus, with more columns, but in practice thirty columns is about the largest number which can be managed. Above this limit, keeping track of the numbers represented becomes too demanding, mentally, for even the most skilful operators.

In any ordinary situation the decision-making process is no problem. The user will be simply repeating a type of calculation continually occurring in his (or more likely, her) daily life. The more important question then arising is, what are the cognitive resources of the proficient user? First and foremost comes a particular conceptualisation of the processes of addition and subtraction, when these are subject to the limitation that there are only four beads below the bar, and only a single bead, representing the number 5 above it. The abacus user cannot simply see $3 + 4$ as 7, since three beads cannot be added to four, to produce this result. In this case 4 must be seen as $5 - 1$, so that $3 + 4$ becomes $3 + [5 - 1]$, and so represents an operation which can be carried out on the abacus, in this case by moving the single bead above the bar down to the central bar, and the lowest of the beads below the bar, down to the frame. That is, adding 4 is an operation requiring one bead to be moved down, both above and below the bar, *provided* of course that the beads can be moved in this way. This will be so if the other number is 1, 2, 3 or 4, but what if it is 0, 5, 6, 7, 8 or 9? With 0 or 5, the answer is simple: the four beads adjacent to the lower frame are all moved up to the centre bar. With 6, 7, 8 or 9, the number 4 must be conceived of as $10 - 5 - 1$, and the sum is carried out by moving one bead back to the frame on both sides of the bar, at the *same time* moving one bead *up* to the centre bar on the column next to the left.[9] This is the operation which makes use of the place-value system. If all this sounds involved, it is second nature to any but a complete tiro.

The procedure for using an abacus for addition differs in two respects from that used in modern western arithmetic, such as it is taught in school or used in practice in commerce.[10] First, the columns of the abacus are worked from left to right; second, every line is added in its entirety, before beginning with the following one: that is, columns of figures are never cast. These two rules may not be essential, but they are well established by practical experience.

Subtraction is simply an inversion of addition, so that the previous paragraph applies with every plus sign being changed to a minus sign, and every minus to a plus, with a corresponding inversion of the roles of the centre bar and the frame. Once again, all this is second nature to any user. This statement is intended to mean that the elementary arithmetical equivalences stated above are only present in the sub-conscious mind of ordinary abacus users. These would no doubt accept the truth of these statements, but find it difficult to see why they need to be made in the first place.

This is an exemplary case of the practical distinction between culture and cognition.[11] The basis of this distinction has been well and simply stated by Lave: 'However one defines cognition, it surely would be located ... in the experiencing of the world and the world experienced, through activity, in context. Culture, on the other hand, is an aspect of the constitutive order' (Lave 1988: 178). One can go further than this: '... activity appears to be routinely efficacious and reflexively intentional without knowing the conditions of its own production' (ibid.: 182).[12]

As far as the day-to-day use of the abacus in Japan is concerned, that puts it in a nutshell. In ninety-nine cases out of a hundred[13] such day-to-day use stops at *kagen*, that is addition or subtraction. If this is the only basic operation capable of being carried out, then the same is true of the digital computer, but just as the computer can be programmed so as to carry out any arithmetical operation, so also can the abacus, and that, essentially, for the same reasons.[14] In practice *jōjo*, that is multiplication and division, is also recognised as an elementary operation on the abacus, but in this case special procedures have to be taught, and particularly for division, a number of alternatives are in current use. Inevitably, both procedures are reduced to processes of addition (in the case of multiplication) and subtraction (in the case of division). The numbers to be added or subtracted, as the case requires, are taken from the multiplication tables for all numbers up to 9. These are just as much second nature to the Japanese abacus user, as they are to anyone in the western world who has finished primary school.

In the case of *jōjo* the procedures followed generally require both numbers to be entered onto the abacus, which must at the same time have column space over for the actual calculations. To take the case of multiplication, which is perhaps more transparent than division, the representation of the first of the two numbers to be multiplied will start with the left-most column of the abacus. It will then take up as many columns as it contains digits (including of course, zero). Leaving a blank space of two columns, the second number will then be entered in the same way. There must then be sufficient columns remaining for the final product to be displayed. If, for example, a three-figure number is to be multiplied by a four-figure number, there must be at least five columns left free at the right-hand end of the abacus, when the calculation is first set up. In this case, then, an abacus with at least fourteen columns will be needed.

The process of multiplication is generally carried out in such a way that the multiplicand is effaced, a column at a time, as it is operated upon, successively, by all the *cyclic* numbers contained in the multiplier. The process is illustrated in figure 9.2. The multiplier, working from left to right, is applied first, to the *right*-hand digit of the multiplicand, which, at the completion of this first step, is effaced. In figure 9.2, which shows the multiplication of 428 by 973, the first step is to multiply 8 by 973, to produce the result, 7,784. This then supplants the digit 8, in the multiplicand, so that the abacus displays, first, 973 (the unchanged multiplier), and then 42 for the remaining two digits of the multiplicand, followed by the result of the first multiplication, 7,784. The second step then applies the same process to the second digit, 2, of the multiplicand, 428, and the third (and final) step applies it to the third digit, 4, reading, in this case, always from right to left. The numbers represented on the abacus, at every stage, are illustrated schematically, in figure 9.2 (where figure 9.1 illustrates the original setting of the beads).[15]

The extension of the use of the abacus to include multiplication and division is but the first stage in a process which can lead to a set of algorithms for any arithmetical procedure. Algorithms for extracting square and cube roots had been developed before the end of the seventeenth century.[16] A remarkable amount of research, sponsored and encouraged by the National Abacus Education Association, is devoted to discovering new algorithms, for use in industry or commerce, and these may be taught in such institutions as commercial and technical colleges,[17] but this is not really the main field of use.

## THE ABACUS IN EVERYDAY LIFE

Historically, the success of the abacus as a calculating instrument is due to the representation of numbers according to a place-value system based on 10. In ordinary market transactions, it did not matter that the calculation effaced its own record, since it was only the final result that mattered. Even today, in the open markets for food to be found throughout Japan, the abacus is used in this way, simply to determine the final total price (*gōkei* or *sōkei*) of the goods sold. The normal use of the abacus is for such *self-liquidating* transactions, so it does not matter that the abacus effaces its own record. The abacus can also be an accessory used for producing a written account using the *kanji* numerals; such usage was certainly quite normal a hundred years ago.

The introduction of Arabic numerals in the latter years of the nineteenth century introduced to the Japanese the possibility of written calculation, in the form that any western child learns in primary school arithmetic. Such calculation, known as *hissan*,[18] is also taught in Japanese schools, and has been in everyday use throughout the twentieth century. The Arabic numerals are not for nothing known as *sanjō sūji*,[19] literally 'calculation use number signs', and this is how the Japanese themselves tend to see them. On the principle that an obsolete technology cannot survive in competition with a more advanced one, the abacus should gradually have fallen out of use in Japan, and at the end of the nineteenth century many expected this to happen.[20] In fact the abacus has not just survived, but flourished in modern Japan.

Until recent research carried out by the National Abacus Education Association,[21] there was little demographic or statistical analysis relating to the distribution and use of the abacus in Japan. Certainly, all Japanese had learnt the basic principles, and could find an abacus to use if need be. An abacus is to be found in almost every home or business in Japan, however modest or pretentious. That the abacus, in its modern Japanese form is an instrument as perfect as the place-value system upon which it is based,[22] is not a sufficient explanation for its popularity.

However perfect the abacus may be, its disadvantages are just as apparent to the Japanese as to any outsider. However great the number of skilled users, the abacus would be quite unable to meet the demands for computing made by the present Japanese economy in the field of finance and accounting, where, historically, its traditional use is to be found. Here the use of computers is no less in Japan than in the rest of the world, where much of the apparatus used is made, or at least designed, in Japan. The economic factors governing the use of such equipment in Japan do not differ from the rest of the modern industrial world, as witness the fact that millions of Japanese never use the abacus at all, even in occupations for which it is well suited. In short, the abacus is not indispensable, and yet it still survives. The reason is not to be found in the formal education system. True instruction in the use of the abacus is part of the curriculum in mathematics for the fourth year of primary school, but the time given to it is not sufficient to train a skilled user. Little is learnt beyond the basic techniques of *kagen*,[23] and it would be safe to say that any Japanese who had not proceeded beyond this level would not use the abacus in everyday life.

All this is radically changed by the fact that something like a half of all Japanese primary school children have abacus lessons out of school-time, in private institutions, known, generically, as *juku*. Many of these, the so-called *kyōiku* (education) *juku* teach other subjects, such as calligraphy or English, but specialised *soroban* (abacus) *juku* are just as common.[24] These come in any number of shapes and sizes, ranging from a converted wooden garage to a small concrete office block, to mention examples to be seen in Kyoto. The courses offered may begin as early as the first year of primary school,[25] and continue on to the sixth or final year, if not longer. The actual programme may depend upon the demands of the parents of the children attending the *juku*, but the contents of the syllabus for every year is more or less uniform, particularly for those *juku* which are members of one of the national organizations.[26] The *juku* aim to achieve standards far higher than the rudimentary level reached by the primary schools, and this is one practical reason why they are so popular. Any *juku* of repute will enter its best students in local if not national competitions, in which the virtuosity achieved by the best performers is simply phenomenal. Even the student who fails to achieve this level, will reach a standard of proficiency equal to that of an expert touch typist in the western world.[27] At this level the use of the abacus is not only very rapid but extremely accurate – two attributes which proficient Japanese users constantly stress, using such words as *hayaku* and *tadashiku.*[28] The result is that in the contexts in which it is used, the abacus does compete successfully with the electronic calculator. When it comes to input, far more mistakes appear to be made with the keys of a calculator than with the beads of any abacus.[29] this may be one reason why commercial high schools also give instruction at an advanced level in the use of the abacus, even when the students also learn to work with the most advanced computers.

The practical advantages of the abacus can be illustrated by its use in the many banks and insurance companies where the processing of figures is a routine operation, now carried out by a computer. The abacus, resting on top of a large computer, is a familiar sight in any such institution. The computer print-out, containing a whole list of figures, can then be checked against the input, which may be a pile of withdrawal slips, to see whether the totals agree with each other. A skilled abacus user can do such a calculation in a matter of seconds.

This is not the main context in which the abacus is used, which is defined by post-offices, open air markets and the countless small shops which are to be found in even the smallest village. In post-offices the sale of different denominations of stamps present something of a challenge, for first the price must be worked out for every denomination by a process of multiplication, and then the total of all these prices must be found by means of simple addition. Although neither arithmetical operation is of itself difficult, their combination demands an algorithm which will utilise the available space on the abacus in an extremely economical way. Few post-office clerks take the trouble to work with such an algorithm; many prefer to perform the multiplication on a pocket computer, or *dentaku*, and then sum the totals on an abacus.[30]

In shops and markets the problem is often solved by displaying different amounts of the goods sold in quantities which are already priced. A stall selling vegetables will have small open baskets containing, for example, onions, in quantities priced at ¥50, ¥100 and ¥200.[31] Other vegetables will be offered and displayed in the same way. The customer then chooses baskets containing the quantities desired of the different vegetables sold, and the stall-holder then empties the contents of all these baskets into the customer's shopping bag, at every step using an abacus to add the price to the existing sub-total. (This may be one reason why the abacus is never used to add columns of figures.) In this case, however, a woman stall-holder, in particular, may not need a real abacus at all, but simply use the fingers of her left hand to move the beads of an imaginary abacus, at the same time as she uses her right hand to fill the shopping basket. This is an example of *anzan*, or mental arithmetic, based on the abacus (Kubo 1986: 160f).

## THE CULT OF THE ABACUS

The earliest recorded use of the abacus in Japan is somewhere around the year 1600 (Toya 1982: 9f). At this time it has already been used in China for many centuries, and its late introduction into Japan can only be explained by the fact that there was no need for it before the growth of commerce in the seventeenth century. Certainly it was always most popular in Osaka, which was to become the great commercial centre of Japan.

With the opening of Japan to the western world, after the Meiji Restoration in 1868, it could have been expected that the abacus would drop out of use, as it had in most of the western world. Such a development would have been a natural result of the education policies of the Meiji governments (Toya 1982: 108f), and it is the continuation of such policies which explains the quite minor part assigned to the abacus in the primary school curriculum of the present day. At the end of the nineteenth century, the official view was that *hissan*, or written calculation based on Arabic numerals, was superior to *shuzan*, or abacus calculation (ibid.: 104). On this view, there was no reason for any official support of instruction in *shuzan*, and even today many young teachers, in particular, would like to see it dropped from the school curriculum.[32] It is easy to argue that Japan, in the age of the fifth-generation computer, does not need the abacus.

Whatever the logic of the argument that the abacus is not indispensable in modern Japan, the members of the supporters' club are still to be counted in millions, and new members are constantly joining them. What is it then that 30 million users see in their chosen instrument (Kubo 1986: 136)?

The answer is to be found in a number of cultural factors relating to the structure of Japanese society. The long-standing Japanese obsession with numbers (which provides the main theme of this book) combines with the popular esteem for physical virtuosity. Jugglers and acrobats are very much part of the popular scene. The abacus, like such performers, is popular in the strict sense of the word, that is, it is an institution of the common people. In social terms there is nothing elitist about the abacus. It is no accident that its most skilled users are often women engaged in petty trade, or sometimes house-wives who use it for domestic accounts.

Seen from this perspective, the art of the abacus, or *shuzandō*, can be compared with *kadō* (flower arrangement), *shodō* (calligraphy) and *chadō* (the tea ceremony) (Kubo 1986: 109). As a pastime any number of merits are attributed to it. It demands dedication (ibid.: 97), accuracy, agility, concentration (ibid.: 113), dexterity (which the Japanese acquire as children when they learn to eat with chopsticks), memory, vision, creativity (ibid.: 168) and, at the highest level, *kyōsōshin*, or the competitive spirit (ibid.: 120). It soothes the nerves (ibid.: 87), and continued use into old age prevents senile dementia (ibid.: 173). All this is in contrast to the mean *dentaku*, which offers no such rewards to its users, who, almost by definition, have no *sūjikankaku*, or feeling for figures (ibid.: 95). It is easy to overlook the point that the representation of numbers, on the abacus, faces in two directions at one and the same time. The one is static, and essentially semiotic: the configuration of the beads on an abacus, at any moment of time, is a structured representation of natural number. The structure at the same time is so elementary as to point to the dynamic direction, which continually creates new structured representations out of the whole according to standard rules. The abacus, in contrast to almost any computer, is no black box.[33] For this reason it provides the easiest possible way to conceptualise numbers, particularly for young children just beginning with arithmetic.[34]

Kubo's anthology covers the whole cult, and places it in a distinctive Japanese setting (Kubo 1986).[35] Its adherents have none the less a lot going for them. If they are proficient, they can compete against other users and attain a *dan*-ranking such as I describe in chapter 10. This provides a built-in incentive to continued progress. The world of the abacus is both popular and hierarchical, which explains, at least in part, its appeal. It is open to all for the cost of a single elementary instrument, which is immune to any threat of obsolescence. The rival world of the electronic computer is elitist in that there is no end in sight to the continued improvements and elaboration of the equipment offered, so that not only capacity but price go far beyond the reach of the man in the street, or better, perhaps, woman in the market-place. The economic muscle of the giant corporation is substituted for the virtuosity of the individual star performer.

If the corporate *sarariman* has no time for the abacus, he is simply out of touch with the vast number of unpretentious Japanese for whom it is part of daily life. The class could include the *sarariman*'s own wife, who may well use an abacus to keep the household accounts. With children, girls more than boys tend to enrol in the *soroban juku*, often simply to join a friend from school. In such a case the motives of the parents are not all that different from those of parents who arrange for their child to have piano lessons, which, if equally popular, are certainly more expensive.[36] In either case there is the additional advantage, important in an overcrowded country, of getting the child out of the home; it is not for nothing that the *juku* have morning sessions during the school vacations.

Finally, in a country where small, tradition-based enterprise has a high status, the continued use of the abacus not only preserves the manufacture of a craft product, but also extends the scope for private enterprise in the field of education. Manufacture, concentrated in two or three provincial centres,[37] is an important source of employment there. At the same time, the capital investment required for a *soroban juku* is extremely modest. The premises of a single shop, furnished with the long low tables to be found in any discount furniture store, is almost all that is needed. The children will sit on the floor as they do at home. There are one or two recognised teaching aids, such as the teacher's giant abacus which can be fixed to the blackboard with suction caps,[38] but these cannot be expensive. With the fees paid by each child, it is not impossible to earn an annual income of more than ¥10,000,000 ($60,000) from a working week consisting of twenty teaching hours. A *soroban juku* is an almost ideal business for a husband and wife to carry on together. The required professional qualifications are no more than can be acquired at any commercial high school. With the present level of demand reaching nearly 50 per cent of children of certain age groups, there is little risk of market saturation.

The result is that many different factors combine to ensure that the abacus not only survives but flourishes in Japan. The abacus may be unsophisticated, but its use is extraordinarily satisfying. The proficient user is a performer, who gains satisfaction which bears little relation to that of, say, a skilled typist in the western world. The proponents of the abacus have been astonishingly creative, in a climate which at first sight is hardly promising. The scope of the abacus, as well as its popularity, are increasing in the world of fifth-generation computers. This may be a popular reaction, but there is now no doubt of the central place of the abacus in Japanese popular culture.

# 10 Games – ancient and modern

## DEFINITION AND TAXONOMY

Games are one kind of play, just as acting is another. Almost any kind of play is based on make-belief, that is, the situation which defines it is contrived and does not of itself relate to the problems of day-to-day living. Almost by definition it is an escape from these problems, and if it has a social purpose, this is what best characterises it. The playing of children is pre-eminently of this kind: it is essentially that part of their life which they are free to live in their own way, to create their own world in opposition to that in which, by force of circumstance, they must grow up. This shows that play is defined also by context, which is almost certain to have both a spatial and a temporal dimension. The wood at the edge of the village, and the hour after school, provide the context for hide and seek,[1] and many other forms of play.

Context has also a social and cultural dimension. In terms of strategy (which is based essentially on arithmetical factors) there is not much to choose between the English bowls and the French *boules*, but the difference in the social and cultural contexts of the two games is obvious to almost any casual observer. The basic activity of a game may itself define two quite different and contrasting contexts in this sense. Golf, at a basic level, consists of hitting a golf ball with a golf club, with maximum effectiveness, but few in Japan would identify those who can afford to play on an actual golf course with the habituees of driving-ranges. Leaving aside the social niceties, and making the distinction in purely systemic terms, one could say that the former are actually playing the *game* of golf, while the latter are not.

The question then is, what makes playing into a game?[2] The line is difficult to draw. Hide and seek is also a game, and yet for any sort of numerical or even logical analysis, it must be excluded from the category. For convenience, then, one must accept that a game is an institution defined by fixed rules. Playing the game *means* obeying the rules (Ahern 1981: 59), and these rules, at least if they are to identify a winner in any instance of the game, must be logical, and their essential basis mathematical. At some stage the game must be reduced to the question, is $x$, greater than, equal to or less than $y$? This may simply be the end result of a numerical system of scoring, such as in the *game* of football. In other cases one must identify the winner and loser by means of a purely binary formula, so that, as in chess competitions, the winner scores 1, the loser, 0, and both players, $1/2$, in the case of a drawn game.

Numbers, in the context of games, are a sort of Pandora's box.[3] Even the most elementary system of scoring allows for almost unlimited elaboration, as one can see from the grand-master points allocated to ranking chess-players. Sometimes such elaboration is built into the game itself: in Japan this is the case with *go*, but not with *shōgi*, the Japanese version of chess. Judged in the light of the rules, scoring may be just one aspect of what makes a game, but it is still necessary to ask how it relates to the rules. To answer this question adequately requires establishing the taxonomy of rules at a deeper level.

There are essentially two cases to consider. The first is where without the rules, the game would not exist. A Japanese example is *go*,[4] but this is but one of many. Rules in this sense are *constitutive*: in principle all that is needed to start playing the game is to read and understand the rules. The whole gamut of board games, such as Monopoly, satisfies this criterion.[5] In many important cases, however, the rules are *regulative*, in that they make a game out of events which can exist apart from any rules. This is what changes riding a horse into horse racing. The cultural transformation involved by this process is almost always quite radical, and tends to place the *game* in a comparative context, in which it can be related to other structurally similar events, in a way impossible at the level of pure play. In this way horse-racing and car-racing come to share common characteristics, not essentially tied to the enjoyment of horses or cars as a leisure activity. The key point is that the essential common denominator is something which can be expressed numerically, which in this instance is speed. With a game depending on regulative rules,[6] their most essential function is to produce the numerical factors, $x$ and $y$, which will determine the outcome.[7] In a game defined by constitutive rules this result is automatic. The significant point is that it follows in either case, something which may be better appreciated in Japan than in the west.

The distinction between the two types of games becomes critical in the field of strategy. This, in a constituted game, must be decided upon by the players on the basic of factors which are essentially numerical.[8] The opposite is true of a regulated game.[9] This distinction is largely effaced in any popular culture, particularly in relation to games, such as baseball or golf, where the rules impose an elaborate numerical framework.[10] This is precisely the reason why, of all the games introduced into Japan from the west in the last hundred or so years, these two are among the most popular.[11]

In classifying games the distinction is often made between skill and chance. This is the basis of a scale whose end-points are defined in terms of 'agon' and 'alea'.[12] The Greek *agon* describes an *open* game, in which the factors determining strategy, at any stage, are apparent to both players. In Japan this is a characteristic of a very wide spectrum of games, ranging from *go* to golf. *Alea*, the Latin for 'dice', refers to the element of pure chance. Although it is possible for a game to be pure *alea*, it is then almost bound to be trivial in any competitive sense. Culturally, however, such a game can be exceedingly important, as the Japanese *janken*, which I consider in the following section, illustrates. The usefulness, however, of the present analysis is to be found in the very large number of games which combine both *agon* and *alea*. In these, *alea* is the factor which governs the arbitrary and uncertain element, such as the identity of the next tile to be drawn in mahjong. The critical point is that where both *agon* and *alea* are present, the former always wins over the latter, at least in the long run.

In a constitutive game, this means that the better mathematician is always the winner. This point, which may be counter-intuitive, is difficult to prove but easy enough to illustrate by means of an example. A mahjong player, when his turn comes to draw a new tile, must also reject a tile already in his rack (which the other players cannot see). In deciding upon his choice, he has a fair amount of information at his disposal, which increases as the game progresses. He knows which tiles have been rejected by the other players, and what open sets they have already formed. This information, at every stage of the game, is always sufficient to determine the optimal strategy. The skill of the winning player consists in being more proficient than the other players in the mathematical processing of this material. The available information is the basis of the *agon* factor; the uncertain order of the tiles still to be drawn from the wall is that of the *alea* factor.

With a regulated game, the position is simpler, in that the skills required are not essentially numerical. Winning at golf is simply a question of being better than one's competitors at hitting the ball so that it reaches a desired point. Even the 24-handicap player has mastered the mathematics of golf; his basic problem is that he cannot hit the ball straight. (This is where *agon* leads to *agony*.)

Strategy determines who wins the actual game, but this, in many important cases, is not the end to the matter. The result of a game, being numerical, can always play a part as a factor in a numerical operation, which in relation to that game is entirely abstract. Such operations serve a number of different purposes, of which three are particularly significant.

The first is to let the result of a single game determine the position of the players (whether individuals or teams) in a competition, *a fortiori* organised on the basis of constitutive rules, designed to produce a winner in the whole field in which the game is played. Although such competitions may be seen as having a distinctive form for each particular game, there is no essential connection between the two, and such forms as the knock-out can be completely general. The socio-economic importance of such competitions, in popular sports, cannot be gainsaid. In the anthropological literature, the classical case is the Balinese cockfight, in which, at the level of the game itself, the protagonists are not even human. But as Geertz points out, 'it is only apparently that cocks are fighting. Actually it is men' (Geertz 1973: 417). At this level, *gambling* [13] on the results – a sort of *meta*-game involving an almost unlimited category of participants – is much more significant than the game itself. For gambling, the game can be dispensed with: any chain of events, which produces numerical results, can be the basis of gambling, [14] although it tends then to be called by another name, such as betting or speculation.

The second purpose served by the numerical results of a game is to rank the players on a recognised scale: in this way a hierarchy of merit is established. This, a notable characteristic of Japanese numerical culture, is considered further in the last section of this chapter.

The third purpose is to analyse the performance of participants, whether individuals or teams. This is almost obsessional with professional baseball commentators on Japanese television, who fill up every free moment with an endless stream of statistics and analysis relating to the players. The same amount of detail is to be found in the sporting pages of the daily newspapers (Whiting 1977: 12), not to mention the magazines devoted to baseball.

These three purposes are clearly interrelated. Ranking is related to performance, and the two combine to create the popular image of the game, which, in turn, is the basis of popular involvement. This is the means by which the people themselves share, vicariously, in the fate of the protagonists (which is what being a fan involves), and gambling extends this commitment into the economic sphere.[15]

This section produces a taxonomy of games in three layers, relating successively to rules, strategy and sub-culture. The rules define the game, whether they do so by *constituting* it, or by *regulating* an established form of play. Strategy is essentially concerned with the process of winning. The sub-culture of a game defines not only its social context and the hierarchy among its players, but also the systemic means by which these are maintained. Every layer has a numerical base, whose importance varies according to each type of game. It may be tempting to build up a typology according to the degree of complexity in every layer, in which case *shōgi*, the Japanese version of chess, would score highly, but a game need not be complex to be significant. The fact that a single game, known to almost all Japanese, is sufficient to make this point, provides the theme for the next section.

## JANKEN

If *janken*, in its basic form a game for two players, is known far outside Japan, it is still entrenched in Japanese culture in a way unparalleled elsewhere. Viewed historically it is but one of many games known, generically, as *ken*, which simply means 'fist'.[16] The explanation is simple. All the different *ken* games depend on both players simultaneously making gestures, in one of a number of prescribed forms, with one or both hands.[17] *Janken* uses only one hand, which can represent three positions, designated as 'scissors' (*hasami*) (index and middle finger apart), 'stone' (*ishi*) (clenched fist) and 'paper' (*kami*) (palm open). The winner is decided on the principle that stone blunts scissors, scissors cut paper, and paper wraps up stone. With two players there is always a result, unless both choose the same position. A drawn position in *ken* is known as *aiko*, but this is no more than a particular instance of the general *shōbu-nashi*, meaning neither victory nor defeat. *Janken* is in fact but one version of the category known as *sansukumi-ken*, literally, 'three-cowering ken' – so that the name is more or less self-explanatory. The difference between different games in this category is in the three positions and what they represent. This difference is not important, so long as the rule of *sansukumi* is adhered to.

Before continuing with the case of *janken* it is useful to place the whole category of *sansukumi* in the general context of *ken*. Historically the first occurrence of *ken* in Japan was with a numbers' game known simply as *honken*, or 'main ken'. The rules are simple. The two players each shout a number from 0 to 10, at the same time displaying any number of fingers of the right hand. If A shouts 'six' and B, 'eight', while A shows two fingers, and B, four, then A is the winner of the round, for $2+4=6$. B then has to pay the penalty of drinking a glass of sake (which of course makes it no great sacrifice to be the loser). *Honken* is also known as *kazuken*, or 'numbers ken', a name which clearly distinguishes it from all forms of *sansukumiken*. *Honken* was essentially a game for convivial social occasions, and as such had its own ritual, which included special, allegedly Chinese words, for the numbers from 0 to 10 (*Daihyakka Jiten* 1951: 8; 387).[18] Originally the different forms of *sansukumiken* occurred in a similar context, with *tōhachiken*[19] being the most popular, but eventually all the different versions, with their social pretensions, fell into decline, leaving the field to *janken*.[20] This, although seen primarily as a children's game by the Japanese, is played in a remarkable number of different contexts, even to a point of being a matter of life and death.[21]

The logical basis of *janken* is defined by a cyclical number-system containing three members. There is, however, an immediate practical obstacle to any purely logical analysis. In no single version of the game are these members designated by a purely abstract symbol, such as a written numeral or a letter of the alphabet. There are three immediate reasons for this defect. The first is quite simply that this sort of logical thinking is out of place in the popular cultural context in which *janken* is played. The second reason is that such abstract symbols lack the necessary metaphorical base which, cognitively speaking, gives the game its essential character. That is, if A, B and C are substituted for *stone*, *scissors* and *paper*, the metaphors based on *blunting*, *cutting* and *wrapping up* are immediately lost. There is no objection to adopting a different instrumentality, based on its own metaphors; this is what explains all the different versions of *sansukumiken*. What cannot be done is dispense with the metaphorical base altogether. Third, once the instrumentality has been decided upon, it operates, at the logical level, as a type of metonymy free from any ordered relationships inherent in any abstract symbolic system. The objection to any A, B, C system is that it contains the implicit relationship that A precedes B, and B, C, while this does not subsist between C and A. Mathematicians are used to overcoming this sort of obstacle when dealing with cyclic groups (simply by ignoring the pre-existing order), but then very few of the players of *janken* have mathematical sophistication at this level.

The mathematics of *janken*, in the context of abstract algebra, is in any case elementary. Much more interesting are the implications of extending the game to more than two players. It is best to start with a simple example, observed on the long flight of steps leading up to the terrace in front of the tomb of the Meiji Emperor in the Kyoto suburb of Momoyama. Three children,[22] starting at the bottom of the flight, are playing *janken*, to see who will first reach the top. In this case, there is only one chance out of three of a 'no win' round. The rest of the time, two children will present the same position, and the third, a different one. The winning child, or children, will then move up a step. If, in theory, one child could move far ahead, or lag far behind, so as to make the game unwieldy, in practice the distribution of winning and losing positions tends to keep the children all within a few steps of each other. In any case, the race up the hundred or so steps will probably take about ten minutes.

In mathematical terms, the general case is defined in terms of an indefinite number of players, $n$. One can conceive of them standing in a circle, with each player making one of the three recognised gestures in every round. Where $n > 2$, the possibility then exists that all three choices will be made, and as $n$ increases, this becomes increasingly probable. The question is, how probable is an outcome, with $n$ players, which divides them into two groups, one of winners, the other of losers. This means that one of the three positions must not occur, nor must the positions of all the players be identical.

For the chance of this happening to be less than one in a hundred, there must be at least fifteen players. If this seems rather a large number, it can happen that *janken* is played on this scale, in certain situations. To take an actual case,[23] fourteen guests at a youth hostel must provide a detail of four to do the washing-up after the evening meal. Many of them are meeting for the first time, but all of them have grown up playing *janken*. Ages range from 15 to 50. The players stand in a circle and start off, making the familiar gestures with their right hands. At this stage the odds are marginally better than one in a hundred that a *decisive* position will occur. Taking a second as the time needed for a single round, there is then a better than even chance of such a position occurring in under a minute,[24] as one round follows another without any interruption.

So it is in this case. In just over half a minute, five players present scissors, and the remaining nine, paper. The five winners are free, leaving the nine losers to go on playing. With the number of players reduced from fourteen to nine, the chance of a *decisive* position is better than one in thirteen, and on the seventh new round, six players present paper and three, a stone. These three are in any case destined to wash up, leaving the remaining six to choose one of their number, by the same process, to join them. This takes them about another five seconds, so that in all, less than a minute proves to be needed to choose a washing-up detail.

Two questions are worth asking. First, what alternative system would be used in a youth hostel in the west? Second, what demands does the Japanese system make on the participants? Strictly, the first question does not need to be answered, but it is almost beyond doubt that western youth-hostellers would take some time to agree a system, and that further, the system eventually agreed upon would need some sort of material accessory, such as a pack of cards. As to the second question, the only demand is a high degree of proficiency in reading *janken* positions among a relative large group of people – something which few outside Japan possess. There is no need to know anything of the underlying probability theory governing the chance of a final outcome after any given number, $N$, of rounds. It is sufficient that for very large $N$ (requiring, say, more three minutes to reach an outcome), the chances are very small indeed, so that they can be treated as negligible.[25]

If, on the face of it, *janken* is rather cumbersome in providing washing-up details, it is very efficient in dividing an even number of players into two teams. All that is needed is for exactly half the players to choose one and the same position, and for this not to be chosen by the other half. The mathematical theory is very involved,[26] but the probability of a *decisive* result is still better than 1 in 20, even where eighteen players have to be divided into two teams of nine for a game of baseball. This is a little awkward, since on every round the majority position has to be counted to see whether it is that of exactly nine players. Playing this system Japanese junior leaguers would still make up two teams in under a minute. With four players, as in golf, there is hardly a problem, since only one round of three fails to produce two pairs. It is hardly surprising, then, that in 1988 a popular magazine published a picture of the prime minister of the day, Nakasone, and three of his cabinet colleagues, on the first tee of a golf-course, each of them presenting a position in *janken*. Choosing partners in this way was second nature to all four of them.

There is no need for citing further instances of *janken*. The lesson from Japan is clear enough. The basic game could hardly be simpler, but its possibilities are endless for a population in which almost every one, above the age of 2, is an adept.

## NUMERICAL STRATEGY

The first section of this chapter has shown how the strategy of games constituted by their rules is essentially numerical, in contrast to that of regulated games. In a short chapter it is, however, impossible to consider, even at the most superficial level, the numerical basis of *strategy* of a single game, let alone all the many constituted games which are popular in Japan. The basis of strategy is the determination of the optimum choice to be made by any player, when it comes to his turn to play. Put in this way, the statement implies that a game consists of a succession of discrete events, each involving one single player. The event becomes complete by the player making his choice and carrying it out. This then sets the stage for the following event.

It is, however, a good question whether the implication stated above is necessarily true for any constituted game. The answer must be that it is. The entire structure of any such game is built up of the discrete units represented by its rules, and at every stage these combine to define the permissible next moves. The numerical result of such a combination is fixed by the interaction of the moves already made and the rules of the game. This is what one sees on a computer screen at any stage of a game, such as *go*, which can be played with a computer. When the next move is made, the previous position is changed by the same process as gave rise to it in the preceding moves of the game. At the same time, every succeeding position, imports any number of possible projections into the future, consisting of the moves yet to be made. Since the basis of all such projections is numerical, the system must, in principle, be capable of ranking them so as to determine which one is optimal. This, ideally, is what determines the strategy, at every stage, of the players, whose turn it is to move. This is the principle on which computer programs for *go*, chess, backgammon, mahjong, or any constituted game, must work.

In practice, the mathematical demands made of an actual flesh and-blood player choosing this approach to strategy are quite insupportable. A chess player would run out of time before he had completed the opening. There is, however, a game of skill, with constituted rules, whose mathematics is *perfectly* manageable by the two players. It is called *nim* and it originated in China. The position at any stage of the game consists of a number of separate piles of some small object, or counter, with no restriction on how many are in any one pile. A player moves by choosing one single pile, and then removing from it as many counters as he wishes. This creates a new position, and the process of reduction ensures that the game comes to an end, with the winning player removing the last pile, within a fixed number of moves. The mathematics requires the number of counters in each pile to be expressed in binary form, which means a series of 1's and 0's. With less than ten piles – in practice, the normal case – these numbers can then be added up, on the decimal system, to produce a total corresponding to the position reached at any time. This number will contain as many digits as there are in the binary representation of the pile or piles with the largest number of counters. If these digits are all even, the position is correct. If both players are competent in the sense of having mastered the mathematical theory, then, at any correct position, the player not having the move must win the game. Since a correct position can always be created out of an incorrect position (the general case of there being at least one odd number in the total), a competent player having the first move must win,[26] unless the opening position is correct. This follows from the utterly transparent nature of the strategy. The result is that *nim* is trivial as a game, if not as a mathematical curiosity.[27]

The point to the discussion about *nim* is that it suggests that the number-theoretical basis of strategy in a non-trivial constituted game is not, and cannot be, what determines the moves chosen by actual players. If it were otherwise, any such game would be no more than a superior instance of the class of games to which *nim* belongs. The most that can be said is that the winning strategies adopted by the top players must be compatible with the mathematical theory, for otherwise a computer program could be designed to beat them. The basis of these strategies in fact transcends such theory, with the result that any useful analysis must be carried out not in the realm of mathematics but in that of cognitive psychology. This is precisely how the Japanese see things. The focus must be upon the way the brain of the master player works, rather than upon any underlying mathematical theory. Once this approach is adopted, it comprehends not only the instant game, but the whole context in which it takes place. On this point Kawabata's *The Master of Go* (1972) is an exemplary case study.[28]

This, however, is not quite an end to the matter. The question remains open as to how far it is necessary to treat the human brain and the computerised theory of games disjunctively. This is tied up with artificial intelligence and fifth-generation computers, matters which I return to in chapter 11. It is sufficient to note here that constituted games must be a focal point of any discussion, and this is certainly true in Japan. In the jargon of the computer age, such games *ideally* represent the *interface* between human cognition and computer software.[29]

How close is this ideal to reality? If Dreyfus and Dreyfus (1986: 24f) are correct, it represents no more than a stage on the road to becoming an expert player. The ideal is presented as such to the beginner, who is taught to think of strategy in terms of rules (such as also define any game), where the lesson from experience is that at some stage the rules must be driven into the unconscious, to be supplanted by a sort of cultivated intuition, if the player is to become a real master of the game.[30] Once this transformation has taken place (and the process may be extremely protracted),[31] the situation confronting, say, a master of *go* whose turn it is to move, is, cognitively speaking, not significantly different from that of a *kendō* expert confronted with the thrust of his opponent's sword[32] (ibid.: 32). If this is so, then in the realm of master strategy, the distinction between constituted and regulated games loses much of its significance.

The analysis so far presented focuses on games with two players. In this case the sequence of play must depend on the players moving alternately, and the only rule necessary is that which determines the player who has the first move. This is generally true of constituted games between two teams. Bridge, which is also popular in Japan, would be an example here. Where there are more than two players, without any division into teams, the rule determining the order of moves is more complicated. The simplest solution is to provide for a cyclical order, and this is the general rule in mahjong.[33] There are, however, situations in which a player can call out of turn, so disrupting the cycle. The interesting point about constituted games for more than two players is that, in contrast to such games as *go* and chess, they are not perfectly open. It is essential to mahjong that the tiles still in the wall are hidden from the players' view. This means that *alea*, the chance element introduced in the first section of this chapter, plays a part, so that in the short run, luck can overturn skill (as any mahjong player knows). This factor does not, however, affect the importance of the role of cultivated intuition. When it comes to strategy, there is no important lesson which cannot be drawn from games with only two players. These also provide the standard model for the Japanese, whether they be *go* or *shōgi* on one side,[34] or *kendō* sword-play or *sumō* wrestling on the other. In the final section to this chapter, all these games and many others will be placed in the distinctive network of competition and ranking, characteristic not only of sport but also of many different arts in Japan.

## SCHOOLS AND RANKING

Japanese sports which are popular in the western world, such as *jūdō* and *aikidō*, which measure competitive success in terms of a *dan* ranking, immediately establish two basic concepts, *dō*,[35] best translated by the English, 'way', and *dan*, literally a 'step', and so, derivatively a 'grade' or 'rank'. Anyone who has climbed a mountain in Japan is likely to have noticed that much of the *way* to the top is marked by *steps*. This is true even of Fuji, where the ten stations, or *gome*, mark the way to the summit.

The leisure activities, recognised as *dō*, are not necessarily competitive sports. The tea ceremony, for instance, belongs to *chadō*, flower arrangement to *kadō*,[36] and calligraphy to *shodō*. Particularly in relation to these activities, which are artistic rather than athletic, the actual organisation of those engaged in them is based on different *schools*, each headed by a particular family, or *iemoto*, whose name it will bear.[37] Within each school the participants will be graded in numbered classes, or *kyū*,[38] with promotion in principle being the prerogative of the *iemoto*. The *kyū* are essentially the grades within the reach of amateurs. At the professional level they are replaced by *dan*, in which promotion is governed by somewhat different criteria.

The combined system of *kyū* and *dan* is characteristic of almost all traditional Japanese sports, from the most gentle intellectual pastimes, such as *go* and *shōgi* to such strenuous physical activities as *aikidō* and the traditional Japanese archery known as *kyūdō*. To take the case of *shōgi*, there are fifteen *kyū* (the lowest being ranked as the fifteenth), and nine *dan* (the highest being ranked as the ninth) (Leggett 1966: 9f). This means that promotion to the first *dan* is the logical next step after attaining the first *kyū*, but any Japanese would see a world of difference between the two. Although amateurs do reach the lower *dan*, none has ever gone higher than the fifth. In the top four *dan* there may be hardly more than a hundred players in the whole of Japan, and those who have attained the ninth *dan* can probably be counted on the fingers of one hand.

Although ascent through the *kyū* may be based on a relatively straightforward points system, a higher *dan* rating is only achieved by a succession of wins against players already at that level.[39] A single victory over a ninth-*dan* player is far from sufficient for achieving the same rank: the process is inevitably drawn out through a long series of games played in the competitive stratosphere.[40] The result is that a position in the highest *dan* is as much the result of longevity as it is of competitive success,[41] so that in *shōgi*, for instance, the real hard fought action between the top players of the day tends to take place at the seventh-*dan* and eighth-*dan* levels.[42]

Traditional Japanese sword fighting, known as *kendō*,[43] is organised on much the same principle, but with six *kyū* and ten *dan*. The *dan*-ratings then provide the basis for three degrees of teacher (*shi*): the lowest of these, *renshi*, is open to anyone above the age of 24, who has been at least three years in the fifth-*dan*. The next degree, *kyōshi*, requires at least seven years as *renshi*, and is open only to those older than 31, while the third, and highest degree, *hanshi*, requires at least twenty years as *kyōshi*, and an age of at least 55. Promotions up to the fifth-*dan* can be made by the local (prefectural) association, or *kendō remmei*, but beyond this they are the prerogative of the *Zen-Nihon* (all Japan) *kendō remmei*. The system, just as in *go* and *shōgi*, works to keep a small number of top players in the public eye, and it is these who control the sport.

The ranking system has its own symbolism, particularly in the martial arts such as *jūdō* and *aikidō*. The traditional loose-fitting clothing, derived from the samurai, is completed by a belt whose colour indicates whether or not a *dan*-ranking has been achieved.[44] In sumo, the names of the highest rank, *yokozuna*, and of the fourth, *komusubi*, both refer to the way the belt is worn.

*Sumō* is worth considering in greater detail for the light it casts on the operation of the principle of hierarchy in traditional Japanese sport.[45] It is a somewhat special case for the reason that a participant seldom maintains his top performance much beyond the age of 30. The hierarchy, like that of league football in the United Kingdom, is based on divisions, subject to the important difference that *sumō* is an individual, and not a team sport. The pyramid is shown in figure 10.1. Only members of the top two divisions, the *maku no uchi*, and below it, the *jūryō*, are actually paid, which is the meaning of the rank of *maegashira*.

In every one of the seven divisions there are six tournaments a year,[46] each with fifteen rounds, spread over fifteen days, and organised so that in every division all the wrestlers will fight each once each day. In each division there are something under forty wrestlers, and the programme is organised so that, so far as possible, every contestant fights those closest to him in the ranking order, both above and below. In any one tournament, therefore, a wrestler will fight approximately half the members of his division. Every day, before each of the main bouts, the participants enter the ring in ascending order of rank, to perform a short ritual of commitment to a pure and fair fight.

*Figure 10.1 Sumō*: the Kanroku scale

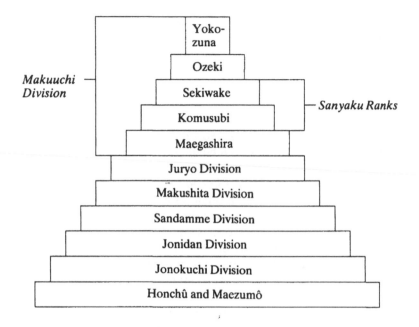

The same order is published in a notice known as a *banzuke*, which lists also the form of the contestants on the basis of previous tournaments. The actual fights on all the fifteen days are not fixed in advance, since in the later days of the tournament the performance of the participants in the earlier days will determine the bouts to be fought. The result is to establish a ranking order at the end of every tournament in close accord with the current performance of each fighter.

At the end of every tournament a wrestler who has won a majority of bouts will be promoted, whereas one who lost a majority of bouts will be demoted, and this will be reflected in the *banzuke* for the following tournament. This, the general rule, does not apply to the two highest ranks of *ōzeki* and *yokozuna*. An *ōzeki* can in principle lose his rank after suffering a majority of losses in two successive tournaments, but in this case would almost certainly prefer retirement. A *yokozuna* cannot be demoted, but will certainly retire as soon he is no longer fit enough to fight in a manner worthy of his rank.

At any one time there are generally only three or four *ōzeki* and no more than a single *yokozuna* (although in the late 1980s there were, for a time, three). The individual contestants in the two next lower ranks, *sekiwake* and *komusubi*, may continually change places, but in the end time will take its toll, and only a few will earn promotion to *ōzeki*, leaving the rest to retire as gracefully as possible before declining powers lead to demotion below the critical rank of *maegashira*.

In the case of *sumō*, the stable, or *heya*, corresponds to the *iemoto*, with at its head the *oyakata*, literally a 'parent' or 'father', who will automatically be a member of the *sumō kyōkai*, the governing body of the sport. The number has been officially fixed at 105 since 1926,[47] so that in anthropological terms the sport is organised on the basis of this number of equal-ranking segments. Every *heya* is divided into two classes, the upper comprising those who, having achieved at least the rank of *maegashira*, earn their keep as competitors, and the lower, the unpaid *tsukebito* or 'helpers', who receive subsistence support from the *oyakata*, and compete in the five lowest divisions. The dividing line will then correspond to that between the upper *dan* and lower *kyū* rankings in a sport such as *shōgi*.

For the anthropologist, the interesting point about the application of a single basic system to ranking in a wide variety of sports and arts is that it creates a pyramidical structure similar to that of Japanese society as a whole. This application of the principle of hierarchy integrates all the participants, from the lowest to the highest, into a single structure. The dominant role of the *iemoto*, in the traditional organisation, reflects the principle of seniority as it applies to the different *ie*, or households, comprising the Japanese village.

At the present time the traditional arts are organised both at local and national level, on the basis of *remmei*, which is best translated as 'league' or 'federation'. There is therefore also a hierarchy of *remmei*, each one being represented at the level immediately above. At the summit, the *remmei* governing the whole country are generally to be identified by a name beginning with *Zen-Nihon*, or 'all Japan', or simply *Kokuritsu*, meaning 'state' or 'nation'.

The *kyū-dan* ranking system is confined to traditional Japanese arts and sports. In many cases it is as much an examination system as a competitive ladder, although even then it may provide the framework in which competition is organised. In the outside world, then, the closest parallel is provided by chess, where players are ranked internationally on the basis of Elo match-points. The system has not been extended to sports such as baseball which Japan has adopted from the west. The competitive structure in such sports, although it may have some distinctive features, is essentially the same as it is outside Japan, although some of the ethos of say, *sumō*, may well be present.

# 11 The ecology of numbers – past and present

## RICE, BUDDHISM AND KANJI

At the end of his preface to the *Kojiki*, the oldest complete work in Japanese literature,[1] Opo nō Yasumarō, who had compiled it from even more ancient texts, signed and dated his work, 'The twenty-eighth day of the first month on the fifth year of Wadō.[2] The Asōmi[3] Opo nō Yasumarō, upper fifth rank and fifth order of merit.' There are five numbers in two not very long sentences. What is interesting is that even today letters and official documents end in much the same way. This evidence of continuity in the use of numbers in Japan leads not only to the question as to how such use arose in the first place, but also to an enquiry into the changes which have occurred in the long course of Japanese history. In a work focused on the present day, the question to ask is, in what way are the changes in the last hundred years important? In this final chapter, the first section is devoted to answering the former question, the second section to answering the latter.

Although many Japanese would find it difficult to believe, there is no reason for supposing that their primordial ancestors had any particular interest in numbers before they first came into contact with Chinese culture sometime around the fourth century AD. Almost every context in which numbers occur, together with the forms in which they occur, can be traced back to Chinese influence. Linguistic evidence, based upon the form of the original, autochthonous numerals,[4] tells nothing about the contexts in which they were used.

Three interrelated factors may be called upon to explain the enormous interest in, and application of numerical methods, from the beginning of the period of Japan's recorded history. Since the records depend entirely upon the use of the characters of written Chinese, known in Japan as *kanji*, this period coincides with that of the pervasive Chinese cultural influence which has continued until the present day.[5]

*Kanji* can be taken as the first of the three factors explaining the traditional Japanese interest in numbers. The reasons are essentially those which I give in chapter 2. There are two different aspects: first there has always been the necessity to impose some sort of lexical organisation on the *kanji* repertoire, and however this is carried out, it is essentially a numerical operation. The *jōyō kanji* now taught to Japanese school children are no more than the latest historical stage. Second, the fact that *kanji*, as a system of signs, gives no special status to numbers, means that they are embedded in the written language in a way which, today at least, has hardly any parallel outside China, Japan and Korea. This establishes numbers pre-eminently as words rather than as abstract, logical constructs. This undoubtedly hindered the use of numbers in arithmetic, and certainly in any advanced mathematics. True, the demands of commerce, in the last 400 years or more, have consistently required efficient arithmetical methods, but this led to these being developed in another and alternative symbolic system – that provided by the beads of the abacus. As I show in chapter 9, the result has been the hypertrophy of one particular numerical technique, so that spectacular results are achieved by the abacus, which have, at the same time, little intrinsic mathematical interest.

Buddhism also came to Japan from China, either directly or by way of Korea, and its essential scriptures became known to the Japanese, *a fortiori*, in their Chinese version. This established a bond between *kanji* and Buddhism, which has no strict historical basis, since Buddhism, in China also, was a religion introduced from outside. The important point here is that Buddhism, at least in the form presented to the Japanese, is a religion rich in number symbolism. This relates particularly to the organisation of time and space. It may be that even before the adoption of Buddhism, ritual events in Japan were already organised by the calendar, but certainly Buddhism, at the very least, greatly extended such organisation. The time-table of mortuary ritual, which I describe in chapter 7, is one important instance of this.

This relates quite directly to the wet-rice cultivation characteristic of Japanese agriculture. If the intricate cultivation cycle, which passes through sowing, transplanting, weeding and flooding to harvest, may not require from the individual farmer a day-to-day reckoning with the passage of time through the seasons, the necessity to co-ordinate the different phases – so as to ensure not only that the supply of water for irrigation is evenly divided between the terraces but also that intensive labour input is available where required – makes the use of a generally accepted calendar indispensable. The time-table cannot simply be decided upon *ad hoc* at the beginning of every new cycle, although in any given year adjustments may need to be made to the fixed calendar to provide for adverse weather conditions, such as drought. Historically Buddhism provides the model of a calendar which determines the dates of ritual events throughout the year. This establishes a model for imposing a fixed time-table upon wet-rice cultivation in any particular locality.

The principle is essentially numerical. Its application can be extended to Shinto by means of a process known as *honji suijaku*, so that rites in local shrines also play a part in regulating the agricultural cycle, at the same time giving the local *kami* a role in assuring a successful harvest. The tie between the ritual calendar and the cultivation is by no means exclusive to Japan. It is to be found throughout the whole area of wet-rice cultivation in South and East Asia, where local populations may be even more attached to number symbolism than the Japanese.[6]

If in the agricultural economy of traditional rural Japan numerical institutions provided the framework for the regulation of life from day to day, no one ever doubted that they could only do so because of a basic accord with the forces of nature. In other words, these institutions were never seen as autonomous; even if their context was religious, they were still derivative from the cosmic order and subject to nature. By force of circumstance, people were conscious of their limitations, and in traditional Japan they were never the basis of any sort of Pythagorean cult of numbers.[7] The whole *samurai* tradition leaned against any such cult, even though such recreations as *wasan* (which I describe in chapter 1) were acceptable.

The conclusion is that although the use and understanding of numbers in modern Japan contains important elements which can be traced back to the beginning of recorded history (and may be popularly regarded as being even more ancient), traditional numeracy is essentially no more than a cultural resource for numerical *bricolage*. The political, economic and above all natural forces which shaped the old tradition no longer operate. Numbers can now be used in ways which are divorced from any historical background. Any Pythagorean tradition, in which numbers are seen as things in their own right, is of recent invention. This is the theme of the final section of this chapter.

## NUMBERS IN THE COMPUTER AGE

In any modern industrial country there is almost unlimited scope for applying numerical lore in everyday life. Some institutions, such as insurance or the stock-market, operate so as to provide numerical information which can guide the policy of those involved in them, according to scientific rules. These institutions, in their actual operation, prove to have their own internal logic. Actuarial mathematics, however much it is focused on one particular context, is still a *logical* system in the strict meaning of the term. Its empirical basis, provided by such statistics as are to be found in mortality tables, is just as firm as that of say, animal genetics or quantum mechanics. The stock-market is a more uncertain case, and yet those active in it still believe in the possibility of *rational* investment,[8] based on the mathematics of the now fashionable econometrics.

Japan is particularly interesting for the firm popular base developed by modern financial institutions. A short walk along Shijō, in the heart of Kyoto's financial district, will pass by a whole row of shops selling stock over the counter. Ordinary people come in from the street and find a place in a row of seats from which they can see banks of television monitors displaying the latest market prices for hundreds of different securities. Any prospective buyer or seller, once he has made up his mind, can go up to a desk and place his order, which will be processed immediately through a computer network. Payment, as likely as not, will be made in cash. It is an open question whether this is any more scientific than consulting the *omikuji* at a local temple, and of course neither activity excludes the other.

The actual technology of over-the-counter dealing in securities is equally available to the individual. After leaving the investment shop on Shijō it is only a short walk to Teramachidōri,[9] where the computer supermarkets offer PC's, known as 'pasocon', or word-processers, known as 'wăpro', in an astonishingly wide range. These are the toys of a numerate population in the modern age. What is significant in Japan is that efficient and successful use depends on finding the right strategies, just as much as in *seimeigaku* (described in chapter 5), consulting the *ekikyō* (described in chapter 6), or even following through the different stages in wet-rice cultivation.

The real transformation is that modern, urban Japanese are freed from the natural restraints operating in the rural economy. The ultimate point in this process of liberation is the game of *pachinko*, where the solitary player has some slight measure of control over a random stream of small metal balls, cascading down through a labyrinth of pins on the vertical face of a glass-fronted console immediately in front of him. Control is confined to turning a single handle which regulates the speed at which the balls are launched into their descent; this in turn determines whether, at the end of the journey, they will be swallowed up by the machine or returned to the player. Skill, such as it is, is confined to a single dimension, and the game becomes an end in itself. For most players, staying ahead of the game means nothing except the privilege of continuing to play without paying for a new supply of balls. Play can in any case continue indefinitely, so long as there is sufficient money in hand to keep the stream of balls flowing.

The remarkable thing about *pachinko* is that it dissolves all categories. It relates to nothing except itself. The individual player, although but one in a low row of identical positions in the *pachinko* salon, is not only quite unrelated to any of the other players around him (who may be numbered in hundreds) but is also involved in an activity which in no way relates to his own past or future (save for the trivial detail that he has just enough money to pay for his supply of balls). Even the obvious numerical basis, which relates the number of balls supplied to a prescribed (but small) sum of money, is lost to consciousness. In terms of time playing at *pachinko* is a sort of mindless *jikan* or space between activities which actually have some purpose. *Pachinko* is a pastime (in the literal sense of the word); it is not a true game.

The truth of the matter is that one cannot be endlessly engaged in solving existential problems. The need to decide, with the help of *seimeigaku*, whether the name of a prospective daughter-in-law is propitious, arises only occasionally. The same is true of the need to decide, with the help of the *ekikyō*, whether a prospective new business venture should be undertaken. At a more mundane level, not everyone is inclined to play such *social* games as *go, shōgi* or mahjong, even if there is someone around to play with. One can look at *sumō* or baseball on television, extracting the maximum numerical potential from the statistics printed in the sporting pages (and flashed, from time to time, on the television screen), but even this becomes problematical for the young man living in a company dormitory or the harassed father whose children want to finish their homework. In such circumstances the driving range or the *pachinko* parlour may be the only escape.

The scope of popular numerical institutions in Japan, both at work and at play, is enormous. The stall-holder in the great Tokyo fish-market of Uogashi, after working all day with an abacus, may stop in for an hour or two's *pachinko* somewhere on the way home to a two-room flat where, with the goodwill of his family, he may be able to look at the evening's baseball on television, with all due attention for the sports commentator's numerical analysis of the progress of the game. What is lacking in such a life style is direction or focus. Numbers are simply the common denominator to quite disparate ways of using time.

The question then arises, are the Japanese today so different in their use of numbers from those who belong to any other advanced industrial civilisation? Baseball, after all, is an import from America, as is the driving range. *Pachinko* is little more than the ultimate over-simplification of electro-mechanical diversion, of a kind to be found is amusement arcades throughout the world.

The abacus, however, is a rather different case. The small electronic calculator, known in Japan as the *dentaku*, is sold in thousands throughout the country. The simplest versions are cheaper than any abacus, and are much more 'user-friendly'. If the abacus is still holding its ground against the *dentaku*, this is because it is a reflection of one sort of socio-economic organisation, in which a whole sector of the economy is dominated by a proliferation of small family businesses. The compulsive users of the abacus are precisely those Japanese who oppose, most passionately, the opening of super-markets owned by multinationals. The abacus becomes, in such a context, a sort of symbol of an alternative economic order, in which the inefficiency of the sector as a whole is somehow overcome by the dexterity of the individuals who work in it.

This is also the context of the 'full name science' of *seimeigaku*, which like the abacus, is seen as an institution of traditional Japan, to be resorted to in solving a traditional Japanese problem. This view is, however, largely illusory. For one thing, the family name in pre-modern Japan was confined to the *samurai*. In the face-to-face society of village Japan there were other, more mundane criteria, for determining matrimonial alliances, and other social or aesthetic factors which determined the right name for a new-born child. *Seimeigaku* is another invented tradition, meeting the needs of a modern, urban society in a culture which takes mass literacy for granted.[10]

The idea behind *seimeigaku* is, however, as ancient as that under-lying the *ekikyō*, or *Book of Changes*. It is that somewhere in the cosmos there are numbers which rule the whole of life. Sometimes, as in the case of the *yakudoshi*, or 'years of fate', they are quite explicit, but in the cases in which *seimeigaku* and *ekikyō* are applicable, they provide the means of gaining access to this realm so as to discover which numbers are critical.

Why then does this ancient understanding of the role of numbers still survive and flourish in modern Japan, to the point even that new traditions, such as *seimeigaku*, need to be invented? Why are there four pages of *ekisha*, or fortune-tellers, listed in the yellow pages of the Kyoto telephone directory, where just *four* names are all that is to be found under the heading 'Astrologers' in the Amsterdam directory? What is it, in modern Japanese culture, which explains the hypertrophy of numerical institutions?

The answer, which I have already suggested in chapter 6, is that a mathematical model is a sort of idealised basis for metaphor. The simplest case – and possibly also the oldest – is the *yin–yang* opposition, with its essential basis in binary arithmetic. The *mandala*, with the same numerical basis, is also of great antiquity, but, as I have already noted in chapter 3, its most remarkable manifestations, such as Borobudur and Angkor Wat, are not in Japan. Implicit in these models is once again 'a unified logic of the universe to be found by conceiving of the reduction of form to number' (Yoshino 1983: 57).

Numbers offer not only logical structure, but also an infinite resource for naming, identifying and interpreting anything with a numerical label, whether it be the licence number of a car[11] or the name of an 800-year old temple. This practice is originally, at least, based on metonymy, in that the number chosen reflects some true attribute of the thing it names. Somewhere around the present-day Tokyo district of Gotanda there was once a rice-field measuring 5 *tan*. The modern age has effaced the need for any such historical connection, which enormously increases the scope for adopting numbers as names. True, the Japanese are not alone in such use of this resource: Vat 69 is not one of the whiskys produced by Suntory,[12] any more than secret agent 007 is a member of the Japanese equivalent of MI5.[13] None the less, the mystique of numbers, if not an exclusive Japanese preserve, is still one which in Japan has been developed to a degree greater than in almost any other country.

The problem, for Japanese culture, may be that the process has now gone too far. Numbers cascade down upon the Japanese, just like the playing cards fell upon Alice, the moment before she awoke from her dream in Wonderland. The real autonomy of numbers does not, in the end, lie in all the metaphors implicit in *janken*, *yin–yang*, *seimeigaku*, but in the austere realm of pure mathematics. It is significant how little purely mathematical properties, such as those of the infinite series of prime numbers, count in the Japanese numbers game, in any of its many different versions. It is in this direction that the Japanese, in the years to come, must deepen their understanding. There is no doubt about the numerical skills of the Japanese: the open question is still about the wisdom of the ways in which they are applied.[14]

# Notes

## 1 THE NUMERICAL PARADOX

1. This intuition is somewhat misleading. Numbers depend upon the creation and use of *unnatural* symbols, but as Lévi-Strauss (1985) has pointed out the symbolism in many traditional societies of interest to anthropologists is rooted in natural phenomena. In such a society there may be little if any concept of number. The point is discussed in greater detail in Crump (1990: 2).
2. In the light of Turing's theorem this is still something of an understatement. The Turing machine reduces not only numbers, but all arithmetical operations, to one and the same elementary code. Its working is described in Bolter (1986: 43f).
3. This is simply the Japanese word for a written Chinese character, which will be used throughout this book.
4. This is a theme constantly reiterated in Hardy (1940), a book written by a leading English mathematician, which is still widely read. This is the world, also, of the Turing machine referred to in note 2.
5. It is worth noting that *arithmos* is the Greek for 'number'.
6. In Japan itself there is a flourishing cult of the indigenous culture, generally known as *nihonjinron*, of which certain manifestations are quite absurd by any scientific criteria. This has, in turn, given rise to a considerable amount of the literature produced by the Japan-bashers among western scholars: see particularly Dale (1986) and Miller (1982).
7. 'Agriculture is an activity in which nature is temporarily tamed or humanised by removing the forest cover' (Gudeman 1986: 6). In the case of peasant cultivators in Panama studied by Gudeman, forest cover was cleared every year, but the principle remains substantially the same.
8. Bray describes wet rice cultivation as '... perhaps the most labour and land intensive cultivation system in the world' (Bray 1986: xv). In China rice has been the most important crop since about AD 800 (ibid.: xiv), and the same is probably true of Japan. The domestication of rice may have begun as early as 5000 BC (ibid.: 9).
9. It is doubtful whether the concept of *wild rice* even exists in Japanese culture. On the other hand, wild grasses are still gathered, even today, to be used for medicinal purposes.
10. Smith (1977: 30–3) is a brief summary of the basis of the taxation of

164    *Notes*

land, paid in rice, in pre-modern Japan. Even when the tax fell to be paid in money, it required 'that rice still be grown in large quantities and sold in order to pay the taxes' (Embree 1939: 113).

11. The mathematical basis of eclipses always defeated the Japanese astronomers (Nakayama 1969: 53), although this did not prevent astrologers from trying to predict them (Crump 1986: 97).

12. See also Gellner: 'The tacit but persistent propaganda by modern philosophy, in quite a broad sense, on behalf of functional specificity, introduced "innocently" as a neutral analytic device in fact favours the mechanistic, disenchanted vision of the world as against magical enchantment.... A sense of the separability and fundamental distinctness of the various functions is the surest way to the disenchantment of the world' (Gellner 1972: 173).

13. True names such as '*alpha* centauri' are still used, but what astronomer today ever sees a centaur in the constellation to which this star belongs?

14. Not every culture commands the use of numbers, as the instances given by Dixon (1980: 107) illustrate.

15. Russell (1920) still remains the best-known popular discussion of this problem.

16. The fact that the ordinal form is in almost every language derived from the cardinal is no proof of the logical priority of the latter: languages simply developed without any regard to this question.

17. This proposition is based on the work of Vygotsky, who has pointed out that the most important event in the life of the young child, generally occurring at about the age of 2, is that 'the development of thought and speech, till then separate, meet and join to initiate a new form of behaviour' (Vygotsky 1962: 43). This new behaviour will include elementary counting at a very early stage.

18. The popularity of the decade in modern western culture is considered in detail in Sontag (1989).

19. This statement is based upon a *scientific* view of the cosmos: the week, defined in terms of the Hebrew *saba'* (seven), is the whole basis of cosmic order established according to the opening chapters of the Book of Genesis.

20. Such terminology is often used as a convenient shorthand, even where it is not strictly necessary. An example is the use of the letters of the Greek alphabet for naming the stars in a constellation according to the order of magnitude. In this case the main purpose is to give convenient names to all stars of interest to astronomers, so that in practice *alpha centauri* need not be identified with the *centaur* constellation.

21. In this case the one set is a permutation of the other. More generally the permutation of ordered sets is essential to many games, particularly all those played with cards, where shuffling is an example of this process.

22. The numeral 4 is standard for 'Thursday' in airline time-tables. In Chinese it is also explicit in the word, *xinggi-si* (lit. weekday-*four*).

23. Mori (1980) is essentially a dictionary of such connotations.

24. It is a good question whether arithmetic must always be so elementary. Hardy and Wright (1945: vii) explain how their book was first to be called 'An introduction to arithmetic', but that this title was abandoned as being capable of misrepresenting its contents.

25. This is also a characteristic of many algebraic systems developed by modern mathematics (Davis and Hersh 1983: 334).
26. Chapter 2 will deal with the question as to the significance of the spoken form of the number, *sanjūsan*, in contrast to its written form, 三十三 .
27. Considering the important mystical connotations of the number 33 in Buddhism, the novelty of its use in the design and naming of the temple of Sanjūsangendō may not have been particularly striking in twelfth century Kyoto.
28. The use of the word 'oriental', instead of 'eastern', is intended to suggest the distinctive but general cultural aspects of that part of the world in which Hinduism or Buddhism have significantly determined the course of history.
29. An example from the English-speaking world is the use of the expression, 'the nineteenth hole', to refer to the bar of a golf club: the special meaning is available simply because a golf course has only eighteen holes. It would be interesting to know if this usage also extended to Japan, where golf is very highly esteemed.
30. Gerschel's (1962) seminal article makes no mention of Japan, but even so, his concept of the *nombre marginal* is one which the Japanese would readily understand.
31. This follows de Saussure, who points out that the succession of sounds in the word has no intrinsic relation with what it refers to (de Saussure 1967: 100). The linguistic *sign* constituted by these sounds is, in the present case, less important than the corresponding visual sign, representing them in written Japanese.
32. The logical basis is discussed in Russell (1920: 64).
33. *Ran* is from *O-ran-du*, which is how Holland is represented in Japanese; *gaku* is a suffix meaning 'learning' or 'science'. By the time of the Meiji Restoration (1868), *rangaku* had been widened to *yōgaku*, or 'western learning' (Sugimoto and Swain 1978: 278).
34. The history of the Dutch in Japan is described in Goodman (1986).
35. The extent of interest at the present is shown by the popular books still published, such as Oya (1980) and Shimodaira (1986).
36. The characters 危険, for *kiken*, provide the common warning for danger, such as comes, for example, from high tension electric power lines.
37. The character 険, common to both *kiken* and *boken*, means steep or inaccessible.

## 2 NUMBERS IN THE WRITTEN AND SPOKEN LANGUAGE

1. This point is discussed further in Crump 1990, chapter 3.
2. There are a few known exceptions to this rule, such as that of the Plateau Tonga of Southern Africa noted by Junod (1927: 166).
3. For this point see Chapter 4, note 6.
4. Although this is not immediately apparent, the words 'eleven' and 'twelve' are composite (Menninger 1977: 83f).
5. In this context, the ordinary mathematical meaning, *not prime*, plainly does not apply.
6. What this involves in the general case is discussed in Hallpike (1979: 216f).

7. In Japan the use of the abacus, examined in chapter 9, illustrates this point.

8. The reader takes for granted that 10 represents the number represented by the word 'ten', to the power of 1, just as 100 represents the same number to the power of 2, and so on, but countless instances from Japan and elsewhere (many of which I discuss in this book) show that this is no more than an assumption implicit in *place-value notation*.

9. The child, on first becoming acquainted with numbers, learns that any number can be reached by counting, and counting to ever higher numbers is part of the means by which it acquires proficiency in their use. At first 'there is no connection between the acquired ability to count and the actual operations of which the child is capable' (Piaget 1952: 61), but then the child comes to realise that the two skills are related, if only to learn at a later stage of development that he must sometimes work with numbers which cannot be reached by counting.

10. *Ichi man* (lit., 'one ten-thousand'), is commonly used, but like in the English 'one hundred' the *one* is strictly otiose.

11. In English, note such words as 'brace' or 'couple' as alternatives for 'two' for use in special contexts.

12. This means that they tend to form part of the vocabulary of the child at a very early stage.

13. English knows no case of *base-suppletion*, that is, the representation of the base by an alternative form, as occurs, for example, in the French *soixante*, where the bound morpheme *-ante*, by means of this process, supplants the base word *dix*, to which it is etymologically quite unrelated. English is also free of inverted forms such as the Dutch *drie en twintig* for 'twenty-three'.

14. A clock striking the hour is hardly an example of speech.

15. This was the principle adopted by the ancient Mayas.

16. This rule leads to difficulties with the number 2, for which there are two versions, *èr* and *liàng* used according to context, each of which has its own distinctive character.

17. The form of the original numerals is given in Ronan and Needham (1981: 2).

18. Strictly *Indian* numerals, as they were correctly referred to in medieval Europe (Murray 1978: 167). It is worth noting that in Arabic, which is written from right to left, the order in which numerals are written is reversed, so that they actually appear in the same form as in the western world.

19. In the terms of the algebraic notation for polynomials this means that the Arabic numeral is a *coefficient* governing an *implicit* power of 10. This point is examined in greater detail in Crump 1978.

20. The use and significance of the abacus, or *soroban*, in modern Japan is the subject matter of chapter 9.

21. The use of the so-called binary numbers occurs in many different cultural contexts in Japan: see e.g. pp. 10 and 35.

22. This is not necessarily true for other languages, even within the Indo-European family. Miller (1986: 51f) takes some trouble to qualify his use of such terms as *noun* and *adjective* in relation to Japanese. Hurford (1987: 157) gets round this point by referring to *nominal modifiers*.

23. In this case it is a good question whether the restriction is entirely due to Chinese, but it is beyond the scope of this book to consider it further.
24. Lit., 'help number word'.
25. For counting up to four *persons*, there are the special autochthonous forms considered on page 24.
26. As chapter 9 will show, this process is quite subconscious for the proficient user.
27. Even so Japanese find it relatively difficult to grasp the distinction between the two sorts of number, particularly when it comes to any sort of logical analysis, such as that presented in chapter 1.
28. The Japanese *dai-ni* (lit. 'second') does not, however, have the negative connotation of the English 'second best'. The *qualitative* use of numbers is less consciously developed in Japan.
29. Chinese goes even further than Japanese, extending this principle to the words for the days of the week.
30. Not to be confused with *ni-ka-getsu*, meaning 'two months', and incorporating the special counter *ka*.
31. Lancy (1983: 142) notes the case of the Ponam language of New Guinea, where ordinal numbers are confined to *first, middle and last*.
32. *Janus* is the god who faces in both directions.
33. The choice of the name *heisei* is explained in Crump (1989: 11).
34. The names, although listed in *kanji*, are ordered according to the place of their component syllables in the *kana* syllabary known as *gojūon* (lit., 'fifty sounds'), which is discussed at the end of this chapter.
35. Historically Yamato represents just one of many different tribes which established themselves in Japan in the face of the original inhabitants, the Ainu (who are now reduced to a small marginal population on the island of Hokkaido). The beginning of Japan as a nation is to be found in the ascendancy which the Yamato succeeded in establishing over their rivals, probably sometime around the beginning of the Christian era.
36. This is presented in detail in Miller 1967 (ch. 2), which is still a somewhat contentious study.
37. See Miller 1971, which explores the possible relationship between Japanese and the other Altaic languages in exhaustive detail.
38. This is particularly the case with *wasan*, which, as chapter 2 shows, was hardly recognised before the sixteenth century.
39. A wide range of *kanji* readings, which are no longer to be found in the ordinary language, still occur in proper names (O'Neill 1972: vii).
40. These alternative forms, being based on the autochthonous numerals, plainly represent a linguistic survival in modern Japanese.
41. *Hatsuka* is an example of a distinctive phenomenon of written Japanese, the so-called *jukujikun*, a compound of two or more *kanji*, whose pronunciation cannot be derived from that of the separate components. To illustrate this point, the first three *kanji* occurring in the written form of 20 December, 十二月二十日, are pronounced according to the normal *on* readings, *jū ni gatsu*, while no break down of the last three *kanji* can show why they should be pronounced *hatsuka*. Another case is provided by 二人, *futari*, for two people. In such cases (of which there are many not involving numbers) the written form often makes transparent what in the spoken is obscure.

42. The complete Old Japanese Binary Numeral System is explained in detail in Miller (1967: 337). The use of designation *binary* is misleading; the system is *decimal*, the binary property being confined to the formulation of even numbers by means of the vowel shift described on page 24.
43. The frame system was not completely standard: different variants for higher powers of 10 are given in Ronan and Needham (1981: 38).
44. This is the number commonly taken to represent the population of Japan, even at the time of the Pacific War, when it was considerably lower (Guillain 1978: 89). The present population (1991) is more than 120,000,000.
45. The reader who really wants to get involved in this question is referred to de Francis 1984, chapter 11, 'The monosyllabic myth'.
46. A small number of *kanji*, known as *kokuji* (lit. 'national character'), are indigenous to Japan, and therefore unknown to Chinese. Surprisingly, though, some of these also have an *on* reading.
47. The fact that the spoken Mandarin for 二, already given in the text, is *èr* and not *ni*, means only that Japanese borrowed the latter from a different spoken version of Chinese.
48. This is how the Japanese themselves see it, but it cannot be taken for granted (in the general non-numerical case) that all the various readings of a given *kanji* are based on some common meaning.
49. There is no rule restricting *kanji* to one reading of either category, although in general there is only one possible *on* reading for expressing numbers above 10.
50. The two syllabaries are both much simplified forms of Chinese characters: the actual derivation is to be found in Miller (1967: 122–3).
51. Terminal *n* only occurs in *on* readings, representing words of Chinese origin.
52. See chapter 3 for a description of the *kan-shi* system of counting in cycles of sixty.
53. In Japan, if you ask someone the way, and you understand the directions given, you can confirm this simply by saying *wakarimasu*; the other person, if unable to help you, could simply and politely use the negative form *wakarimasen*.
54. Japanese contains many other words and usages represented by 分, but they are not important in the present context.
55. *No*, a particle which constantly occurs in Japanese, has the effect, in the present context, of turning the word before it into a genitive, like the English *'s*. A literal translation would then be *of six parts, five*.
56. Unger (1987) not only explains the complexities of written Japanese, but examines their consequences for modern computer science.
57. This principle, if fundamental in Japanese cognition, is disputed by western scholars. See, for instance, de Francis' extreme claim that 'there never has been, and never can be, such a thing as an ideographic system of writing' (de Francis 1984: 133).
58. In medieval Europe the cultural significance of numbers was much closer to that of the orient (Crump 1990: 46).
59. Some *Kanji* have alternative versions, each of which may have a different number of strokes. There are thirteen different kinds of stroke, each with its own name, but there is no rule preventing a stroke being repeated, so

that $\Xi$, for the number, 3, consists simply of the horizontal stroke (*ōkaku*) written 3 times.

60. The list developed out of an earlier one of the so-called *tōyō kanji* which was adopted as a temporary expedient just after the end of the Pacific War. The transformation of the *tōyō* into the *jōyō kanji* is a lesson in Japanese political manoeuvering, well described in Unger (1987: 92–8), but hardly relevant in the present context. For practical purposes the differences between the two lists are about as subtle as the difference in meaning between *tōyō* and *jōyō*, which may both be translated into English as 'everyday use'.

61. The subject matter of the whole chapter should make it clear that the dictionary is one which explains the meaning of Japanese words to Japanese readers.

62. *A fortiori*, since the whole system comes from China in the first place.

63. The child is precocious since 寿 is only learnt in the *third* year of primary school.


## 3 ALTERNATIVE NUMBER SYSTEMS

1. According to Russell, the specific 'question, "What is a number ?" ... has only been correctly answered in our own time. The answer was given by Frege in 1884, in his *Grundlagen der Arithmetik*' (Russell 1920: 11). The implications of the *infinite* series of cardinal numbers must have been known at least in the time of Euclid, when the existence of an infinity of prime numbers was first proved (Hardy 1940: 32–3).

2. The reader of a book will be inclined to take for granted, perhaps subconsciously, that this representation is visual. The cognitive implications of this assumption are made beautifully clear by a short passage from *Alice Through the Looking Glass*:

> 'You don't know what you're talking about!' cried Humpty Dumpty. 'How many days are there in a year?'
> 'Three hundred and sixty-five,' said Alice.
> 'And how many birthdays have you?'
> 'One.'
> 'And if you take one from three hundred and sixty-five, what remains?'
> 'Three hundred and sixty-four, of course.'
> Humpty Dumpty looked doubtful. 'I'd rather see that done on paper,' he said.
> Alice couldn't help smiling as she took out her memorandum book, and worked the sum for him:
>
> $$\begin{array}{r} 365 \\ -\ \ 1 \\ \hline 364 \end{array}$$
>
> Humpty Dumpty took the book and looked at it carefully. 'That seems to be done right –' he began (Gardner 1970: 267–8).

Lewis Carroll leaves it to the reader to note that the transformation from the spoken to the written form represents a cognitive shift from one numerical system (based on the constantly repeated annual cycle of 365 days) to another, that of the infinite series of cardinal numbers.

3. In modern semiology many, such as Jakobson, would question the extreme position taken by Humpty Dumpty (Barthes 1967: 21).

4. The fifty-two cards in a pack have almost as many meanings as there are card games for which they are used. For a general discussion of the numerical systems used in games see chapter 10.

5. Modern Chinese has simplified these two characters to the forms 阳 and 阴, where the *non-radical* elements, 月 and 日, are the familiar characters for *moon* and *sun*. Japanese, however, still uses the original *kanji* forms, 陰 and 陽.

6. Modern mathematics is certainly capable of devising alternative systems for ordering the infinite series of natural numbers, but they lead one into very deep waters. For a general discussion see Davis and Hersh (1983: 152f).

7. *Dō* is of course cognate with the Mandarin *dào*, and as such is an *on* reading of the Chinese character; the alternative *kun* reading, *michi*, is the ordinary Japanese word for a street or road.

8. Ronan and Needham's postscript is worth quoting *in extenso*:

> In his correspondence with the Jesuits in Peking in the seventeenth century ... Leibniz brought himself into contact with the *I Ching*, and as he examined the long and short lines of the hexagrams, he came to see what seemed an astonishing fact – the lines could be identified with binary numbers, counting in a scale of two (the type of counting used in computers and electronic calculators), which he had himself invented a few years earlier in place of the usual scale of ten. The Chinese seemed to him to have discovered the system many centuries before. However, the opinion of Leibniz was not quite valid: the men who had invented the hexagrams were simply concerned with all the permutations and combinations they could make from their two basic elements, the long and the short strokes.

9. This is a form of fortune-telling known in Japan as *ekikyō* (which I describe in chapter 6), and in China as *I Ching* (under which name it has become popular in the west).

10. The use of the word 'algorithm' is intended to indicate any recognised logical process, which can be applied in appropriate contexts to achieve a specific result.

11. There are any number of references to the *mandala* in the literature of oriental cultures, but a good short analysis is to be found in Tambiah (1976: 102), while Snodgrass (1985) relates it to one of its most important manifestations, the *stupa*, which can be seen as the most important component in Buddhist symbolism. For a short general discussion see Crump (1990: 70–1).

12. The word 'topological' is chosen to make clear that it is the relationship of the components of the structure to each other which counts, and not its measurable dimensions.

13. This formula is noted by Shorto who has observed it occurring 'time and again in political contexts in South-East Asia' (Shorto 1963: 581).

14. The system of decimal (or *Arabic*) numerals also provides instances of

marginal numbers in this form: see, for instance, the thousand and one nights of Sharazad (whose life was saved because she succeeded in living one night longer than the number of nights which could be counted).

15. The fact that the core has but a single unit, with every successive container having an even number must mean that the total number of units is always odd.

16. Most odd numbers, even if relatively low, have never been represented in a *mandala*; it is doubtful whether anyone has ever taken the trouble to compile a list of those which are represented. For an interesting special case (not from Japan) see Shorto (1963).

17. The problem with three is that a core in a container consisting of but two poles is somewhat exiguous to constitute a recognisable *mandala*. More generally the number of units comprised in any *mandala* is likely to be of the form 4n + 1 rather than 4n + 3, for this allows every successive container to be symmetrical about both axes. The simplest form of a *mandala* is thus represented by a compass with the four cardinal points.

18. The *locus classicus* must be the thirty-seven *nats* of the Burmese pantheon, for which a complex *mandala*, derived from the thirty-two *myos*, or townships of the medieval Mon kingdom of Lower Burma, provides the blueprint (Shorto 1963).

19. Yoshino (1983: 32) lists some of the possibilities recognised in Japan: a much longer list, for China, is given in Ronan and Needham (1978: 154–5).

20. This is confusing, because *chi*, as in *tenchi*, is also earth. But *do* (土) is essentially the element earth, where *chi* (地) is earth as opposed to heaven. These are both *on* readings; both *kanji* have the *kun* reading, *tsuchi*.

21. The days of the week in Japanese, starting with Sunday, are named after the sun and moon, followed by the five planets in the order Mars, Mercury, Jupiter, Venus and Saturn. The seven-day week is, however, a comparatively recent innovation, and never part of the basic Chinese measurement of time, adopted in Japan.

22. The mathematics is surprisingly involved. If the number of elements in the cycle is not prime, some elementary transformations will cause decomposition into a number of smaller cycles, identical to each other. Even if this is avoided, alternation between two possible configurations, although it sometimes occurs, is not the general case.

23. Needham lists four *enumeration orders*, and of these, the mutual production order is equivalent to *sōshō*, and the mutual conquest order, to *sōsoku*. Assuming the equivalence of mirror images and a cyclic order for the five elements, there are twelve possible *enumeration orders*. These in turn divide into six pairs, with every pair having the same transformation property as that given in figure 3.2 between *sōshō* and *sōsoku*.

24. In this case the sixth month was taken to be under the sign of the earth.

25. Needham (1980: 126) illustrates this by means of a diagram representing a sine wave, which is reproduced in Crump (1990: 53).

26. The full name science of *seimeigaku*, which I describe in chapter 5, incorporates both systems for its own idiosyncratic purposes.

27. Together they form, therefore, the *jikkan jūnishi*.

28. These are listed in Palmer (1986: 35–6).

29. *Ka* is the *on* reading of 歌, so that *tanka* is a 'short song' and a *waka*, a 'wa-

song', with *wa* having the same meaning of *autochthonous Japanese* as in *wasan* and *wagaku.* On a strict historical analysis the two are not quite equivalent.

30. The alternative *kun* reading, *nagauta,* also occurs.
31. Compare the institution of *omikuji* discussed in chapter 6.

## 4  THE CULTURE OF NUMBERS

1. Literally, 'a bird in the hand is better than ten in the air'. The change of setting is no more important than the change in numerical value.
2. Consider the line from an English hymn, 'A thousand ages in thy sight are but a moment gone'. If 'a thousand ages' were replaced by 'two moments' the message would be banal, to say the least.
3. It could be argued here that statistical considerations must restrict $b$ to a relatively small number. Beyond a certain critical threshold, it would obviously be worthwhile to take one's chances with the birds in the bush.
4. Matthew ii. i.
5. Kaspar, Balthasar and Melchior: these do not occur earlier than the eighth century.
6. According to Euclid, 'a number is an aggregate composed of units' so that 1 is not a number. This position was certainly accepted throughout the middle ages (Menninger 1977: 19).
7. This is the final refrain of every stanza of *Green Grow the Rushes O.*
8. Here *hitori* is an instance of *jukujikan,* a word whose meaning can only be analysed according to the written form. See chapter 2, note 41.
9. Note also the expression *koku ikkoku,* or 'moment by moment', or, more commonly in English, 'hour by hour'.
10. Mori (1980: 44) cites both the *Kojiki* and *Nihonshoki,* the two epic narratives of the mythical history of Japan.
11. The book of Isaiah (xlv. 7) makes clear that God created all things, both good and evil. For the Japanese position see Moeran (1985: 92f).
12. It is understood here that it is the leaf of the *aogiri* or Chinese parasol tree.
13. A variant of the same theme occurs in *ippan wo mite zenhyō wo shiru,* or 'see one spot and one knows the whole leopard'.
14. In English, compare 'too many cooks spoil the broth' with 'many hands make light work', noting in both cases the implicit numerical base. See also note 3 above.
15. One is reminded of the couplet from Robert Browning's *By the Fire-side*:

> Oh, the little more, and how much it is!
> And the little less, and what worlds away!

16. This is essentially the argument that the Japanese put forward for using *kanji* in the written language.
17. The alternative *on* reading, *nishin,* of the same two *kanji,* also occurs with the same meaning.
18. The use of the *on* reading, *ni,* is probably more normal than that of the *kun* reading, *futa,* in the two examples already given.
19. The expression *nike,* literally 'two hairs', points the contrast between black and white horses, suggesting at the same time of the possibility of a mixed colour, brown (Mori 1980: 55).

20. Compare the New Testament narrative of Mary and Martha (Luke x. 38–42), which imports a moral judgement not found in Buddhism.
21. Compare the *sanpō*, the three treasures of Buddhism – the Buddha, the Sutras and the priesthood.
22. There is an alternative *sangai*, whose meaning is almost identical. The two substantives, *se* and *kai* combine to form *sekai*, the ordinary Japanese word for 'world'.
23. Note also in chapter 8 the Buddhist significance of 33.
24. The word used here, *kuku*, repeats the root for suffering, to connote double suffering.
25. This is the suffering of Paolo and Francesco in Dante's Inferno (v. 121):

> Nessun maggior dolore
> Che ricordarsi del tempo felice
> Nella miseria.

26. This is the suffering of Job.
27. *Shiki* is also used, with special connotations in *onyōdō*.
28. But compare the quite different connotations of *ji*, given above in relation to the number three.
29. In this latter case there are also the *rokujō*, or six emotions, but these are not constituted simply by adding one extra to the *gojō*.
30. These figure in a rather cryptic proverb: *gohon no yubi de kiru nimo kirarenu*, literally, 'in cutting with five fingers, no cut is made', the implication being that no one finger is better to be cut than any other. This imports a metaphor referring to the members of a family, so that it is impossible to chose one as a scape-goat for a shared misfortune. This is a typical dilemma in the traditional rural society of Japan.
31. A Japanese league (*ri*) is about two and a half miles.
32. Chapter 8 describes how the names given to the historic wards of Kyoto extend the number of directions to 9, by including not only the centre, but also left and right.
33. Compare the *shihōkai*, based on the number 4 and also connoting the universe, which comes from the tradition of Kegon, a Buddhist sect originating in eighth century Japan.
34. Other words for a pawnshop are *shichiten* and *shichishō*, *Shichi*, in its meaning of *pawn*, is a prefix to a dozen or so associated compound words.
35. The number 8 occurs far more than any other *small* number in the *Kojiki*, so that its mystical significance almost certainly predates any Chinese influence. This relates arithmetically to the fact that 8 is the third power of 2, and the last such power beneath the critical threshold of 10. For the *Kojiki* see particularly Philippi (1968: 80ff).
36. Compare the Christian doctrine from John xix. 2: 'In my Father's house are many mansions'.
37. I am grateful to Prof. Toshinao Yoneyama of the University of Kyoto for this case. Another case of unfavourable connotations of 8 is the expression *hatsugo no kanshaku*, or 'eight child tantrums', which described the anger of an 8-year-old child whose protestations go unheeded.
38. This is very involved, since the rules applied also combine with the *gogyō* and *ekikyō*, described in chapter 6. Similar complex principles govern *seimeigaku*, described in chapter 5.

39. These are the four cardinal points, and the four intermediate points, plus the centre. As one would expect for heaven, the model is that of a simple *mandala*, as defined in chapter 3.

40. Is there a Japanese equivalent of 'a cat has nine lives'? Kanyō Kotowaza Jiten (1988) lists at least a dozen proverbs relating to cats, but only one of these has any numerical content: *Neko wa sannen no on wo mikka de wasureru*, meaning, 'In three days a cat forgets three years' kindness'. Dogs in Japan are more highly thought of, for *inu wa mikka kaeba sannen on wo wasurenu*, meaning, 'A dog which is fed for three days will not in three years forget the favour'.

41. The phrase *kyūgyū ichimō*, which is based on an identical paradigm, has a quite different meaning. The literal translation is 'nine cows one hair', so that not surprisingly the meaning is much the same as the English 'a drop in a bucket'.

42. These are examined in detail in chapter 2.

43. *Jūichibun*, literally 'eleven parts', and *Jūnibun*, 'twelve parts', can both mean *more than enough*. Mori (1980: 391) gives only the former, and Nelson (1974: 214), only the latter meaning.

44. This is an example of a type of word known as *jukujikun*, of which the component *kanji* of the written form cannot be related to the syllables of the word. In the present case, therefore, it is only the written form of *mukade* which supports the analysis given in the text. *Yuri*, meaning a 'lily', is another example of a *jukujikun* based on a hundred. In Japan, incidentally, *mukade* are poisonous.

45. *Tsuzura* is another *jukujikun*. *Ori* is the first part of the familiar *origami*.

46. The component *ka* in these two words have different meanings, represented by different *kanji*. The former means 'faculty' or 'subject', the latter 'goods for sale'.

47. See chapter 5, pages 66–7.

48. *Yao*, for 800, is archaic, and only survives in a few special uses, such as those cited in the text. It is however common in the oldest Japanese literature, such as the *Kojiki* (Philippi 1968: 85f). The normal word for 800 is *happyaku*.

49. In Mori (1980: 3) or Nelson (1974: 943) compare the poverty of the entries under *rei*, with the wealth of those relating to *man*. The Japanese have no problems with simple negation, such as indicated, in countless compounds, by the suffix *mu-*.

50. It occurs, for instance, in the title of Okubo (1978). It would be interesting to know when the Japanese population was first considered large enough to be measured in this way. Guillain (1979) records that it was current before the Pacific War, at a time when there were only 65,000,000 Japanese.

51. There is a pun in this proverb. The written form uses 金, meaning 'gold', but the Japanese pound, written, 斤, has the same spoken form, *kin*.

52. Much the same meaning is to be found in the expression *ichi ji no shi*, a 'teacher of one character'. This is a mark of great esteem, the principle being that a good master can show how a single character, correctly used, can determine the entire meaning of the passage in which it occurs.

53. The fact that it means 'ninety' excludes the use of *kujū* for 'nine or ten'.

54. These are the liver, lungs, heart, kidneys and spleen.

55. These are the colon, the small intestine, gall bladder, stomach, entrails and bladder.
56. Note also that *hara* belongs to the men's (yang) speech, where women (yin) use *onuka*: Ohnuki-Tierney (1984: 59). This explains why ritual suicide in the form of *harakiri* (or in the more common, reverse, *on* reading, *seppuku*), is a male preserve.
57. The difference between *kyō* and *shū* is not easy to translate into English. They refer equally to faith, belief or doctrine, perhaps in a rather formal sense. They combine to form *shūkyō*, which is the normal translation of 'religion'. *Kyō* is also the Japanese for 'sutra', and the basis of *kegon* is the *kegonkyō*, which the Chinese, Tao Hsuan, brought to Japan in the year 736.
58. The Buddhist calendar relating to death is considered in detail in chapter 7.
59. Lit., 'initial *nanoka*'.
60. A *ryō* is an old Japanese coin, no longer in current use.
61. This form is not listed in the *Concise Oxford Dictionary of Proverbs*, which gives, 'An ounce of practice is worth a pound of precept', together with a number of variants.
62. According to the *Concise Oxford Dictionary*, '*ship* is a dialectical pronunciation of *sheep*'. This makes sense when one realises that tar was used to protect sores against flies.
63. Although the character is based on the *wood* radical, its use is by no means restricted to structures made out of wood.
64. Is there a parallel here in the spoken language in that *hyaku* (100) less *t* (1) becomes *haku* (white)?

## 5 WHAT'S IN A JAPANESE NAME?

1. The tickets issued for train reservations now take the form of a computer print-out with the place-names written in *katakana*, but this is a concession to the need for relatively simple apparatus. Even so the input by the ticket clerk is by means of keys labelled in *kanji*, although arranged in the standard *kana* order of the *gojūon*, as described in chapter 2. This is, of course, by no means the only example of such usage. The development of word-processing programmes to work with *kanji* is a part of Japanese computer science with no parallel in the western world.
2. In Japan it is taken for granted that every character has its own meaning (Toyota 1984: Preface), but see also chapter 2, note 57.
3. For the non-Japanese reader this makes O'Neill (1972) absolutely indispensable.
4. Although a Japanese equivalent is perfectly possible, none is listed by O'Neill (1972).
5. The popular cognitive basis of numbers, and number use, is the subject matter of Lave (1988). This study makes clear, however, that such use is much more culture-specific than mathematicians and educationalists allow for (ibid.: 173).
6. Among English place names, Sevenoaks and perhaps the much more modest Six Mile Bottom, spring to mind. Forty Hill, in north London, is, however, not related etymologically to the number 40. Given names such

as Tertius or Octavia sound exotic, and examples of surnames, such as Twentyman, are very rare.

7. Pre-modern Japan was divided into *kuni*, or provinces; *kuni* is the *kun* reading of the character, 国, whose *on* reading is *koku*. In Shikoku (lit. four-*kuni*) the four original provinces have been replaced by four prefectures (*ken*) with different boundaries. Kyushu confusingly, contained eleven, and not nine, *kuni*, the *shū* representing a different political unit, of which there were nine in the island.

8. It is fourth only in population; in area it is the second largest island, although it only comprises a single prefecture, out of the sixty-six in Japan as a whole.

9. This rule may have been established in China as early as the fourth century BC (Ronan and Needham 1981: 187).

10. In England, also, there is no *south country* corresponding to the current *north country*.

11. This is still the official name for the present capital city of Japan: it is the form commonly used for written addresses, but it is hardly ever heard in ordinary conversation.

12. The form *kyō-to*, comprises the *on* readings of the *kanji* 京都: these have the same *kun* reading, *miyako*, but all readings have the common meaning of 'capital city'.

13. Mikuni also occurs as family name, in the *kanji* form, 三九二, meaning simply 3-9-2 (O'Neill 1972: 262).

14. Compare the Dreiländerspitze, a mountain in the Alps, at the meeting point of three of the old Austrian provinces.

15. Tokyo, at the time it became the capital of Japan, was known familiarly as 'the city of 808 streets' (Fukuzawa 1966: 197).

16. The *kanji* 子, in its *kun* reading, *ko*, almost invariably indicates a girl's name.

17. Strictly, for reasons given in chapter 2 relating to the use of counters, or *josūshi*, the correct Japanese expression would be *futari no kodomo*, which is listed in Mori (1980: 57). The form in the Tokyo place name represents an archaic usage.

18. The Japanese normally talk of *suiden*, where *sui* means 'water' and *den* is the *on* equivalent to *da*. This term is plainly appropriate to wet-rice cultivation. In practice *da* occurs most commonly in countless Japanese surnames like Tanaka and Uchida.

19. Once again, because of the rules governing the use of *josūshi* referred to in note 17 above, the correct form should be 'roppon *no* ki', corresponding to the name Sanbon *no* Hashira (*Three Pillars*) known to Japanese art and literature (O'Neill 1972: 293).

20. O'Neill (1972) lists only *kudan*, and even this does not appear in Mori (1980).

21. One is reminded of John Buchan's famous novel, *The Thirty-nine Steps*, where 39 was the actual number of steps, which confirmed the discovery, at the end of the book, of the secret upon which the whole action turned.

22. Compare *banzai*, literally 'ten thousand years', the traditional Japanese cry of congratulation.

23. See Crump (1989: 100f) for a description of the *daijōsai*, the ancient 'great new food festival', which completes the installation of a new

emperor. In November 1990 this festival was celebrated for the new emperor, Akihito, almost exactly sixty-two years after it had been celebrated for his father, the Showa Emperor.

24. The other is *Mie* (lit., 'threefold').

25. In Japanese, *hantō*, literally 'half island', like the German *halbinsel.* O'Neill (1972: 41) lists only one instance of *han* (in the sense of 'half' occurring in a place name, and this, not surprisingly, is Handa, or 'half a rice-field' (which also occurs as a surname).

26. Literally, 'great names'. The *daimyō* were the 300 or so members of the highest order of the nobility, who were directly responsible to the *shōgun* for the government of their fiefs.

27. *Kanji* compounds, when not representing a proper noun, tend to be *on* readings, often for abstract nouns. It can well happen that such a compound, in its *kun* reading is a name, and in its *on* reading, a noun. In principle there is no reason why the two should not both occur in the same sentence, in which case only context would indicate the correct reading.

28. Compare note 16 above.

29. Based on the number 7. Girls' names ending in *ko* are based also on 2, 5, 8 and 9. By analogy with such familiar names as Mitsui and Mitsubishi, based on 3, one would suppose that this was true also of Mitsuko, the given name of the famous pianist, Uchida Mitsuko. This is, however, not the case, for Mitsuko is based on a non-standard reading of 光, meaning 'light'.

30. The normal reading is *bu* as in *bushidō*, the 'way of the samurai'.

31. Although some names are written in *hiragana*, this is not at all usual. It follows that the meaning of a name would have to be looked up on the basis of the component *kanji*, as given in a standard dictionary. This is often extremely problematic. Names are generally compounds, but not in a form which otherwise has a recognised meaning. The only possible meaning, then, will require a conflation of two disparate components, often used with readings which deviate from standard Japanese. What these readings will then mean, particularly in combination with each other, is by no means easy to fathom.

32. Other instances are given in chapters 6 and 7.

33. According to normal practice the family name precedes the given name.

34. See chapter 3, pages 38–41.

35. This can always be looked up in a dictionary, but most adult Japanese can, with some difficulty, recall the *on* readings out of their own memory. This process used to be essential for using Japanese typewriters, in which the order of the characters was determined according to the *on* readings. Now, with the word-processors described in chapter 11, this arduous process is no longer necessary.

36. This conclusion is supported by the compounds listed for the *kanji* for both words listed in Nelson (1974: 194 and 265f).

37. This, the general case, is subject to the last full paragraph on page 67. Here, also, it is worth noting that the given name of Admiral Yamamoto (who in December 1941 launched the Japanese attack on Pearl Harbor) was *Isoroku,* a deviant reading of 五十六, meaning simply '56'. This name was chosen by his father who, at that advanced age, was proud to have a new son.

38. In a table based upon the sum of two or more numbers, 1 can never occur. It is, however, theoretically possible for a name to consist of but a single one-stroke *kanji.* An example would be Kinoto, 乙, which occurs in the *kanshi* table (fig. 3.3, page 42). This is listed as a place name in O'Neill (1972: 3), which also lists an alternative reading, Todomu, as a boy's name.

39. *Kichi* is the *on* reading equivalent to the *kun* reading, *yoshi.*

40. A fifth case, *nakashibari,* or 'tied to the centre', can arise when the full name comprises five or six *kanji.* In this case the four outside signs are the same, as must be the two inside signs where there are six *kanji.*

41. This is the first part of *bushidō,* the name given to the samurai code of chivalry. This explains why the *kun* reading, Take[shi], is so auspicious.

42. Murō (1983) is a Japanese presentation of Risshō Kōseikai.

43. The significance of Nichiren for sectarian Buddhism is described in MacFarland (1967: 180f). I discuss in chapter 7 the numerical basis of the Lotus Sutra, which is for the Nichiren sects the most important of the Buddhist scriptures.

44. The publisher is listed as Kaiun Shinbunsha: significantly *kaiun* means 'improvement of one's fortune'.

45. This is more significant than appears on the face of things, since it is a shinto priest who performs all the rituals for a new-born child (Picken 1980: 25). I examine the place of Shinto and Buddhism in the Japanese life cycle in chapter 7.

46. The rules are much more complex, but they ensure that the practice of using given names as a form of address is extremely restricted. Among grown men, even intimate friends will not use given names, although they may well use the affectionate form, *chan* after the family name. The whole position is explained in detail in Yano (1973).

47. As Berque points out, 'in Japanese, "I" is a succession of terms (*watakushi, boku, ore* etc.) which are topologically determined by the environment (Berque 1986: 102). The use of the word, *topologically,* is misleading, since it disguises the fact that the environment is primarily social.

48. The use of the word 'demonstrative' here is philosophical rather than linguistic. Ayer (1963: 148) lists four demonstratives in this sense, *this, I, here* and *now.* In the Japanese context I deal with this point in Crump (1988).

49. As Wittgenstein points out, the meaning of a name does not die with the death of its bearer. If it were otherwise ... 'it would make no sense to say "Mr. N.N. is dead"' (Wittgenstein 1958: 20). The Japanese translation (Fujimoto and Sakai (1968: 254) of the whole passage containing this excerpt fails to preserve Wittgenstein's distinction between *meaning* and *sense* (using *imi* in both cases), and renders 'nonsense' with a neologism, *nansensu.* In view of the large number of reprints in the last twenty-odd years, *The Philosophical Investigations* obviously appeals to the Japanese, but given the linguistic problems (which I consider in Crump 1988) it is difficult to see what they can make of it. In the particular case of proper names, Wittgenstein is certainly in tune with Japanese thought.

50. These names are known as either *hōmyō* or *kaimyō,* according to the branch of Buddhism looking after the mortuary rites. My own enquiries

at the Higashi Honganji Temple in Kyoto suggest that the numerical rules of *seimeigaku* play no part in the selection process.
51. In Crump (1989: 11) I describe how bureaucratic requirements decided the choice of Heisei as the name of the era of the present emperor.

## 6 FORTUNE-TELLING

1. The normal word for 'calculation' is *kei-san*; here *kei* implies measurement, which is no part of *uranai*.
2. As with any Japanese noun, the form *omikuji* is apt for both singular and plural. Although logically its use, in Japanese, is plural, I have chosen to treat it, for the purposes of English grammar, as singular.
3. Compare *o-mi-ki*, 御神酒, for the wine offerings made as part of Shinto ritual.
4. For the rare individuals born before 1892 who are still interested in their horoscope, the chart can be extended by simple extrapolation. The number 99 has no particular significance, except for its being a convenient multiple of 9.
5. The code letter T (for *tree*) is chosen to avoid confusion with the element, water.
6. See figure 3.3, page 42.
7. Strictly the Emperor Meiji died in 1912, but this is counted as the first year of his successor, the Emperor Taishō. The same correction must be made also for the latter's death, and that of his son, the Emperor Shōwa.
8. The table is not included in the almanac for 1990.
9. The almanac does not give any reason for the variations in the order of the numbers for different years with the same *kishō*.
10. This is printed on the same page of the 1983 almanac.
11. In practice any Japanese would know the animal for the year of his birth without having to look it up in an almanac.
12. Since the divination forming the basis of the almanac is derived from the lunar calendar, in which the year begins with the festival of *setsubun* (marking the end of winter) at the beginning of February, those whose birthdays occur before then must be regarded as having been born in the old year.
13. I explain the traditional Chinese reasons for this orientation in Crump (1989: 97f).
14. This undoubtedly follows an earlier Chinese legend, in which a mythical beast, half dragon and half horse, emerged from the Yellow River, with the numbers from 1 to 10 inscribed on its back.
15. All the squares in figure 6.6 are derived from the centre square, which is the only magic square, with 5 at its core. The following squares are found by adding 1, successively, to every one of the nine numbers in this square. The addition is carried out (mod 9), so that 9 is subtracted from any number greater than 9. The nine squares in figure 6.6 exhaust this process.
16. Because of the occurrence of leap-years it cannot be said that the same sequence of *kanshi* repeats itself every sixty years, any more than it can be said that every seventh year begins on the same day of the week.
17. Yoshino (1983: 60) refers to this, suggestively, as *tatami no me*, literally,

'the fold between two *tatami* mats'. On this point see also chapter 8.

18. The former is the direction from *yin* to *yang* the latter from *yang* to *yin* (Yoshino 1983: 60).

19. According to figure 6.6, 1 June (the date of writing this note), being nine days earlier should also be *kichi*, where in fact it is given as *hanhan* in the 1990 almanac. The answer is that the rule of figure 6.6 is sometimes changed by one step, for no apparent reason.

20. The Korean flag contains four of the eight *hakke* placed at the intermediate cardinal points around a core consisting of the yin–yang emblem. It need hardly be said that Japanese and Korean divination have much in common. For the latter see Janelli (1977).

21. Blofeld (1976: 216–19) gives, and explains, a number of possible orders of the *hakke.*

22. The mathematical relationship between the numbers of the lines and that of the corresponding hexagram is too intricate to be explained in the present text.

23. See note 13 above.

24. This is said only to give some idea of the length; in the Japanese of the *ekikyō* it is hardly appropriate to speak of the number of *lines.*

25. *Eki-ha kono shō wo kazu ni kangen shite shikō shi, sore ni yotte uchū ni okeru tōitsuteki no ri wo motomete iru.*

26. The whole principle is clearly stated in a dialogue contained in Lodge:

> Metaphor is a figure of speech based on similarity, whereas metonymy is based on contiguity. In metaphor you substitute something *like* the thing you mean for the thing itself, whereas in metonymy you substitute some attribute or cause or effect of the thing for the thing itself. (Lodge 1989: 222)

For further elucidation of this point, see ibid.: 178, 336f.

27. If, for instance, in such a model, *i*, represents the rate of interest, its mathematical behaviour within the model can actually correspond to the fluctuation of the rate of interest in any actual situation. This was how the model was constructed in the first place. There cannot be any corresponding process in a divination model.

28. The Japanese folly is well described in Tuchman (1984: 34f).

## 7  TIME

1. This point is made by Caillet (1986: 45) with the reference to the French *quand.* Among the examples given in *The Kodansha English–Japanese Dictionary* are 'ame ga furu *toki* niwa ...' ('when it rains ...' but literally 'rain falling time in') and 'tegami o kakiageru *toki* ni ...' ('when you finish the letter ...' but literally 'letter finished time at'). There are also any number of contexts in which the translation of *when* does not involve *toki.*

2. The more normal word for 'sun' is *taiyō*, literally 'great *yang*', where moon is *taiin*, 'great *yin*'.

3. The *kun* readings, *aida*, and less commonly, *ma*, can also mean an *interval.*

4. Note that the two words combine to form the basis of *kantensen*, a 'dotted line'.

5. According to Caillet (1986: 45) *jikan* literally means 'intervals between *toki*'. In this sense *toki* is eventful time, where *jikan* is empty time. Certainly in the domain of space, the equivalent to *jikan* is *kūkan*, which means simply 'space', and is in no sense a unit of measurement.

6. For the spatial connotations of *aida* see Berque (1982: 63).

7. For another related use of *setsu* see note 18.

8. The *kanji* 周期 shows how this word combines the circularity of *shū* with the expectation of *ki*.

9. Once again this is a case of *kan* 間 defining an interval of time as in *jikan*, 'hour'. Given the historical background, 週 must be a neologism, if the term can be applied to a newly created *kanji*. The question is, why were the Japanese not content to use 周, also read as *shū*, to mean 'week' in the compound *shūkan*? The answer may be found in the connotation of 'advancing' along a 'road' in the radical ⻌ of the *kanji* 週.

10. The Chinese *fen* is also the verb, 'divide', but then in Chinese there is no possible distinction between *on* and *kun* readings.

11. The same principle was followed in determining the dates of the monthly examinations (Fukuzawa 1966: 81).

12. There are also rites performed in the course of pregnancy. The most important of these occurs on the day of the dog in the fifth month, when the mother first wears the special sash, or *iwata obi*. At the present time the consultant obstetrician writes the *kanji* for 'happiness' on the *obi*, after which a nurse shows how it is to be tied (Ohnuki-Tierney 1984: 182f).

13. Compare *haka-mairi*, the normal word for the visits to the tomb prescribed for the 14th, 21st, 35th, 49th and 100th days after death, and then on the 3rd, 7th, 13th, 17th, 23rd, 27th, 33rd, 37th, 50th, and sometimes even the 100th anniversaries.

14. Traditionally two different periods of mourning were also prescribed, one governing the wearing of mourning clothes, and the other (which was always much shorter), abstinence from animal food. The length of time prescribed depended on the mourners' relationship with the deceased. As one would expect, the longest periods were required for a husband (but not a wife) and parents (including adopted parents). The actual periods are given in Chamberlain (1974: 337).

15. This is the present state of affairs. Before the adoption of the western calendar, *bon* was a midsummer festival occurring on the seventh day of the seventh month.

16. Reckoned from the old-style new year at the beginning of February, this puts the period of storms in September, which in fact is when typhoons are most likely.

17. Because Islam, which also adopted the lunar month, failed to provide for this adjustment, the period of *ramadan* occurs ten days earlier in every year.

18. The basis was a solar year divided into twelve *setsu* of equal duration. The midpoint of each *setsu* was the *chūki*, and the winter solstice was the first *chūki*. (*Ki* and *setsu* combine to form *kisetsu*, the normal Japanese word for 'season'.) Since the interval between two successive *chūki* was

slightly longer than the lunar month, it could happen, occasionally, that such a month contained no *chūki*, in which case it was deemed to be an intercalary (*jun* or *urū*) month, with the same number as that preceding it. This occurred approximately seven times every nineteen years, but over the long term all twelve lunar months would be repeated as an intercalary month.

19. There is of course some system in the juxtaposition of certain stars and planets, the return of comets or the occurrence of eclipses, all of which have long been known to the Japanese. To find a repetitive order in such events that can be used to count years as the phases of the moon can count months is, however, extremely difficult.

20. For the historical background and numerical significance, see chapter 8.

21. In February 1989, the charge for these services at the Tsurugaoka Hachiman Shrine in Kamakura was ¥3,000. A typical shrine scene is illustrated in Yoneyama (1990: 174), showing a notice board with the words *yaku yoke oharai*, or 'ill fortune protection purification', but not mentioning the price to be paid.

22. A complete list is given in Lewis (1986: 169).

23. Norbeck (1955: 118) rejects such 'folk explanations', where Lewis (1986: 170), who examines the whole question in some detail, is inclined to accept them.

24. But see Ogawa and Grant (1990) for the way in which *yakudoshi* have been transformed in Hawaii to provide the occasion for a party – a custom now adopted by many Americans not of Japanese descent.

25. The period is thirty-five days (also a multiple of seven) in the case of a woman. Smith (1974: 92f) gives the Buddhist origins of rites determined by intervals occurring in multiples of seven.

26. This is the historical explanation for funeral ritual being a Buddhist preserve. It is significant that when, at the end of the nineteenth century, Shinto funerals were first allowed (Chamberlain (1974: 204), the Buddhist succession of seven-day rites was replaced by five 10-day rites, culminating on the fiftieth day after death (Smith 1974: 74).

27. The alternative policy, *Shinbutsu bunri*, of separating Buddhism from Shinto, only became important after the Meiji Restoration of 1868, when it was essential for establishing the position of the emperor in state Shinto (Smith 1974: 28).

28. Miyako (1980: 19) simply refers to *this world* and *that (the other) world.*

29. This transformation is also expressed in terms of the opposition between *haka*, the grave marked with the name of the departed, and *yama*, the mountains which are the home of the gods. According to Smith, 'one of the oldest and most widespread beliefs as to the final destination of the soul is that it resides in mountains' (Smith 1974: 63). In this connection see also Smith's discussion of the double-grave system (*ryō-bosei*) (ibid.: 75). The connotation of *mountain* is also explicit in the word *sanryō* for the graves of the last two emperors (Shōwa and Taishō) for whom the mortuary rites were pure Shinto.

30. For a detailed description of the domestic celebration of Bon see Smith (1974: 99).

31. Embree (1939: 294) displays a chart of the lunar calendar year, in its relation to the agricultural cycle, in which three busy seasons alternate with three slack seasons.

32. As Embree (1939: 295) noted more than fifty years ago, the lunar cycle is gradually being broken up and old holidays are losing meaning. At the same time it is to be noted that the two equinoxes, by definition belonging to the solar calendar, were, in Japanese Buddhism, always important ritual events, known as *higan*. Once again the dead are commemorated with a visit to clean the grave and services in the home or local temple (Smith 1974: 99). In the Meiji era *higan* was also celebrated in Shinto as a spring festival of the imperial ancestors (Chamberlain 1974: 159).

33. In principle Ise should be as old as the imperial line, officially claimed to have been founded in 660 BC, on 11 February, which is still the day of a public holiday known as *kigensetsu*. According to the historical record the shrine was only designated as the home of the sun goddess at the end of the seventh century AD (Smith 1974: 8), and in fact the last reconstruction, in 1973, was no more than the sixtieth. The 1953 reconstruction should have taken place in 1949, but this was not possible under the allied occupation of Japan, which therefore had the effect of retarding the cycle by four years.

34. It does not matter that Islam is based on a lunar month, although this does mean that the sacred month of *ramadan* occurs some twelve days earlier in every year.

35. The Christian and Islamic calendars, being based on historical events, then have to solve the problem of numbering the years before these events. The solution, essentially, is to use negative numbers, which in the Christian calendar are indicated, parenthetically, by 'before Christ' (BC). This causes no problem to historians, who tend not to be concerned by the logical and cognitive problems inherent in this solution.

36. This is also the orthodox Jewish solution.

37. This was hardly a problem for the Japanese intelligentsia. Yukichi Fuku-zawa (1966), who lived through eight *nengō* before he was 30, never had any difficulty with dates.

38. In England the official state calendar, still used for dating legislation, is based on the number of years of the reigning monarch. The normal calendar, in which the current year is 1990, was originally that of the church, in opposition to the state.

39. This has the somewhat paradoxical consequence that in a ceremony known as *tsuigō hōkoku no gi* and performed on the twenty-seventh day after death, the deceased emperor is ritually informed of his *new* name, which is identical to that known to him during his life as his *nengō* (Tanaka 1988: 264).

40. During his life the emperor was hardly ever referred to by name in Japan. The designation *tennō* was sufficient, for at any one time there could never be more than one emperor.

41. These include the chronicles of *Kojiki* and *Nihonshoki* and the *Ekikyō*, the Japanese version of the *I Ching*, or *Book of Changes*. I discuss the calendar processes of the Japanese court in greater detail in Crump 1989, chapter 1.

42. The two others were ruled out because, written in *rōmaji*, they began with S, the same initial letter as in Shōwa. Computerised data-banks had already listed dates from 1926 (the year of the Shōwa Emperor's accession) with the prefix S, so that another letter was required for the *nengō*

of the new emperor. No one has so far come to grips with the problem of what will happen when all the twenty-six letters of the *rōmaji* alphabet have been used up by the time of the accession of some remote future emperor. The Japanese will no doubt cross that bridge when they come to it (as they undoubtedly will if the past historical record is any guide).

43. Its only basis are the classic texts of the *Kojiki* and *Nihonshoki*, written more than a thousand years later, which are the earliest indigenous Japanese texts (and even then the *Nihonshoki* is written in Chinese).

44. This is still a public holiday, even though the Allied Occupation of Japan (1945–51) suppressed the cult of the emperor. See also note 31.

45. Grew (1945) gives a full description of these events by the American ambassador.

46. The word is used advisedly: this was very much a case of a tradition 'actually invented, constructed and formally instituted' (Hobsbawm and Ranger 1983: 1).

47. The *kanshi* numbers reflect the ideals of Pythagoras, who was concerned to 'establish a numerical cosmology on the basis of the non-mathematical associations of numbers' (Crump 1986: 89). The presence of such associations is essential to the use of the *kanshi* in horoscopes, which I discussed in chapter 6.

48. I was not yet born in 1920, nor do I expect to be alive in 2054.

49. For the first time in Japanese history the reign of the Shōwa Emperor left room for possible ambiguity, since it lasted long enough for the *kanshi* cycle to repeat itself. However only four years, *heiin, teibō, boshin* and *koshi*, were repeated, before the emperor died in January 1989.

50. Only once in Chinese history did the system break down with a reign of more than sixty years, that of the Emperor K'ang Hsi (1662–1723). With the beginning of the republic in 1912 China faced greater difficulties. This year was treated as the beginning of a new era, which in Taiwan still continues even though some thirty years have now repeated themselves. Hong Kong abandoned the system in 1948 for diplomatic reasons.

51. Even towards the end of the 1980s, when the Shōwa Emperor was approaching his ninetieth birthday, it was not difficult to find notice boards relating to public works reading 'due for completion in Shōwa 66' or some other future year which would never actually occur, seeing that the emperor was to die in the first week of Shōwa 64.

52. The traditional Japanese day had twelve hours, of which six belonged to the day and six to the night, so that in winter the former were shorter than the latter, and in the summer, longer. According to the twelve *shi*, the first hour was that in which midnight was the mid-point, equivalent to 11 p.m. to 1 a.m. This was also the ninth hour of the night, which, somewhat confusingly, would be followed by the eighth hour, and so on until the fourth hour, which would be that immediately before, but not including, midday. Midday would then once again be the ninth, and the succession would repeat itself until the fourth hour, which would then be that immediately before, but not including, midnight. These hours are named according to the autochthonous (*kun*) readings of the written *kanji* numerals, for which see page 26.

53. These, in the order in which they occur in the days of the Japanese week, are Mars, Mercury, Jupiter, Venus and Saturn. As such they correspond

to the five elements, fire, water, wood, metal and earth, which provide the Japanese names for these five planets.

54. Compare the camera which now automatically records the date on every picture taken, which I consider in chapter 8.
55. For the latest trial run of the French *train à grande vitesse*, S > 500. One may be certain that the Japanese will equal this record.
56. The italics are mine in the passages cited from Caillet (1986).
57. It is interesting to note the view of Yukichi Fukuzawa, the most progressive intellectual force in nineteenth-century Japan, who was himself born a lower *samurai*:

> In the education of the East, so often saturated with Confucian teaching, I find two things lacking; that is to say, a lack of studies in number and reason in material culture, and a lack of the idea of independence in spiritual culture.... I believe no one can escape the laws of number and reason, nor can anyone depend on anything but the doctrine of independence as long as nations exist and mankind is to thrive. Japan could not assert herself among the great nations of the world without full recognition and practice of these two principles. (Fukuzawa 1966: 215)

## 8 THE SPATIAL WORLD OF NUMBERS

1. The negative, inchoate, connotations of *sora*, should be contrasted with the positive, ordered connotations of *ten* (天 not 点), generally translated 'heaven', but with a wealth of other meanings.
2. It is however important if mathematical graph theory is to be applied to the analysis of social networks, as I have done (for an Indian community in southern Mexico) in Crump (1980).
3. Other more complicated constructs (e.g. figures of 8), such as occur in graph theory, are quite marginal to the present analysis.
4. An alternative here would be the Tokyo *Yama no te*, which like the Osaka *kanjōsen* runs above ground. Both have any number of junctions and intersections with other lines.
5. The figures come from the current *jikokuhyō* (see page 97) of the Japanese railways.
6. In the Kyoto system the sixth and seventh stations, travelling south from the northern terminus at Kitaōji, are numbered '4' and '5' according to the East–West streets, Shijō and Gojō, intersected by the line. On this point see also page 125 (for Kyoto street names). On the Osaka *kanjōsen* the name of only one station, Nishi*kujō*, incorporates the number 9.
7. In the time-table Tennōji, an important junction in the south of the city, is taken as the starting point, simply because it is where the first trains of the day, in either direction, start.
8. At first sight it would seem that this result was achieved more than a thousand years ago with the *thousand day pilgrimage on Mount Hiei*, which originated some time in the ninth century (Reader 1985: 9). The actual pilgrimage is divided into ten periods of 100 days, spread over seven years, and although prescribed routes must be followed in the

different periods, their actual length (which can be considerable) is not *numerically* significant.

9. The train announcements make the point clear. 'The 1105 *shinkansen* for Tokyo is now arriving at platform 11' is all the information that the impatient *sarariman* wants to hear. This train actually has its own name, not based on the time-table, *Hikari 136*, but in the context this is much less informative.

10. For the implications of this at different stages of Japanese history see Ohnuki-Tierney 1989, chapter 4.

11. Significantly *uchi* is also a normal word for 'house'.

12. Masai (1987: 69) suggests that where a European city is always on a river, a Japanese city must be close to a mountain. The point is significant, for a mountain is a point (*ten*) where a river is a line (*sen*).

13. At the present day the achievement is becoming more and more unimpressive. In 1980, when I climbed Fuji, a bus took me to the *gogōme*, or the fifth stage, so that my staff only records the five higher stages. As for the sacred Mount Ontake described by Blacker (1986, ch. 14), it is now possible to reach the top by cable car.

14. In the case of Ontake there are by necessary implications from Blacker (1986: 293) at least two alternative routes.

15. Compare the way in which Islamic pilgrims circle the black stone, or *kaaba*, at Mecca (Turner 1974: 177).

16. Turner does not appear to allow for this possibility at all, even though he notes that the Arabic *hadj* is probably derived from an old Semitic root, *h-dj*, meaning 'to go around, to go in a circle' (Turner 1974: 173). In this context see also Leach's (1966: 125) distinction between the two aspects of time, discussed in chapter 7 above.

17. There are also a number of affiliated temples, which are not strictly part of the pilgrimage. These are known as *bangai*, literally 'ex-number'.

18. This is a posthumous name. The founder during his life was known as Kūkai.

19. This is the basis of the distinction which the Japanese make between the Saikoku and Shikoku pilgrimages. The former focus on a particular figure, such as the Kannon Buddha, to whom each temple is dedicated, whereas the latter focus on a charismatic figure who accompanies the pilgrim along his route (Reader 1988: 51).

20. Miyazaki (1985: 38–42), noting the significance of 33 in the *saikoku* pilgrimage, still fails to find any convincing explanation for 88, although he considers a number of possibilities. Many of these recall instances occurring elsewhere in this book.

21. Reader (1988: 59) tells of one route less than a kilometre long, in which the Shikoku temples are represented by small stones and statues.

22. This took Dr Reader and his wife forty days in the late winter of 1984.

23. All that is necessary is to get a stamp or seal (*hōin*) from every temple. Sometimes this is left to the bus-driver, while the passengers remain in their seats without even visiting the temple. The whole cultural transformation is described in Reader (1987b).

24. Turner's essay makes scarcely any use of Japanese material.

25. On a visit to the island of Utsukushima in the inland sea in 1984, I took the cable car to the top of the mountain, with the object of descending by

the path. A Japanese, whom I had not noticed before, asked to accompany me, so that at every station on the path, I could take his photograph. So doing, I provided the *hōin* necessary to authenticate his descent. Later, in 1987, a visit to the Olympus head office in Shinjuku (Tokyo) confirmed that Japanese insisted on cameras which recorded the date of every picture, while there was no demand for them in the overseas market.

26. Note, however, the archery contest in the gallery running along the back (west side) of the pavilion, described in chapter 10.

27. As such it developed from, and replaced a form of building known as *sentai kannondō*, literally '1,000 bodies *kannon* hall'.

28. The name connotes both the purity of the Buddha, and his capacity to overlook the failings of mankind.

29. The Kannon Buddha is a perfect *bodhisattva*, characterised by boundless compassion and mercy. The Lotus Sutra (Jap. *hokekyō*), which is central in Japanese Buddhism, is the most important scripture relating to the Kannon Buddha.

30. Similar Buddhas are to be found in seventeen of the temples of the *sanjūsankasho*.

31. In relation to the wooden buildings of northern Europe, which at the dawn of the middle ages provided the modular basis of Gothic architecture, see Horn (1975).

32. For some 600 years the *tatami* has consisted of a base (*toko*) of woven straw, covered with a soft surface (*omote*) of woven rush. The short edges have coloured cloth borders (*herinuno* – originally indicating the owner's rank), which enable them to be sewn together. The fact that the name comes from *tatamu*, to 'fold', suggests that the mats in their original form could be folded and stored. It is only relatively recently that mass production has reduced the price sufficiently for *tatami* to be the normal floor-covering in Japanese dwellings.

33. Where necessary a square, half-*tatami*, can be used.

34. To give one example, the extensive complex of buildings comprising the Tsurugaoka Hachiman shrine in Kamakura (as illustrated in Picken 1980: 81–2) lies at the end of a road leading almost due south to the sea, at Yuigahama beach, about a mile away. The main shrine buildings are surrounded by a square corridor, which is placed symmetrically at the end of the approach road. The number of bays on each side of the corridor is almost certainly significant.

35. The system is explained in Pye (1977: 64–5). In the instance given in the text, Fūjiyama is a division (known as a *chō* or *machi*) of Yamashina, one of eleven wards (*ku*) of Kyoto, which corresponds to the postal district 607. Strictly speaking 607 Fūjiyama-chō 2-182 would be a sufficient address, but in practice any Japanese would write Kyoto(shi) 607, Yamashina-ku, Fūjiyama-chō 2-182, followed by the name of the addressee. This corresponds to an analytical process which starts with the largest unit, the city or prefecture, and works down to the smallest, the individual – the reverse of the normal western practice.

36. In this way *Sanjō* is the street by which the old Tokaidō enters the city. There is also a *Hyakumandōri*, or 'Million Street', also running east-west, but its name is not part of any ordinal number system.

37. The word is sometimes translated ward, although this is confusing in the case of Kyoto, where the districts, or *machi* into which the old city centre is divided, are also translated by ward. The Kyoto *machi* are still important for their separate contributions to the Gion Matsuri, the festival celebrated by the people of Kyoto, on 23 July, over a period of more than a thousand years (Yoneyama 1974: 26–7). More generally, there is a whole hierarchy of units, of successively decreasing size, into which the whole of Japan is divided. The largest is the *ken*, or prefecture, which consists of rural districts, *gun*, and cities, *shi*. The *ku* is the next largest division of a *shi*, followed by *chō, chō-me, banchi* and *gō*, the number identifying each separate plot of land.
38. Literally 'East *mountain*', referring to the range of hills, with at the northern end the sacred Mount Hiei, on the eastern side of the city.
39. The present analysis comprises only nine wards, when there are in fact eleven. The two remaining wards, Fushimi and Yamashina, were not, however, part of the original city.
40. The traditional court of the emperor was organised on the same basis, so that there were ministers of the left and ministers of the right.

## 9 THE JAPANESE ABACUS

1. See for instance Ronan and Needham (1981: 35), where it is suggested that the origin of the word 'abacus' is to be found in the Semitic *abq*. This is interesting, because 'calculation' is certainly derived from the Latin *calculus*, meaning a pebble. Seen in this light, calculation with the abacus can be seen as originating with moving *pebbles* over *dust* (or *sand*). Smith notes that the abacus may have been known in Babylon, on the basis of a character in Babylonian cuneiform apparently derived from a representation of the abacus (Smith 1958: 40).
2. It is natural to take for granted that a numerical system has but a single base, such as the number 10 in the decimal system. This was true of almost all Chinese systems of measurement, but an abacus could well be designed to deal with special cases, such as the pounds, shillings and pence of the British monetary system before 1971. Japanese money, at the time the abacus came into use, was also based on a decimal system, in which one yen was equal to a hundred sen. This was only partly true of other systems of weights and measures, as recorded, for instance, by Nelson (1974: 1029f).
3. The abacus may be said to combine the merits of the Chinese numerals (the ideal spoken system) and those of the Arabic numerals (the ideal written system). I discuss this point further in Crump (1990: 45f).
4. It can also be zero, in which case the number represented would simply be 753.
5. This can be avoided simply by marking, at the top of every column, its place in the frame system; any number of abacuses have been produced with such markings in the form of the Chinese numerals of the frame system. The giant abacuses used by Japanese school-teachers in front of the class are often marked with the *kanji* numerals shown in figure 3.1. The advantages this has in teaching the use of the abacus at elementary

level, are, however, soon outweighed by the disadvantages following from the loss of flexibility.

6. The setting could equally represent the single number 97,300,428, multiplied, *a fortiori*, by any power of 10.

7. In principle the forefinger is always used for moving the single (= 5) bead above the bar, and also for moving any bead below the bar back to the frame. The thumb is then only used for moving any such bead up to the bar. The basis of this principle is purely ergonomic, and any user is free to deviate from it if it suits him.

8. Johnston (1985: 215) contains a fascinating discussion of this point in its relation to the Mandarin bureaucracy in China.

9. If in this column all four beads are already at the centre bar, then 1 has to be conceived of as $5 - 4$, or $10 - 5 - 4$; in the latter case, the calculation has to be followed through to the then next column to the left, and so on until it can be completed within a single column.

10. This statement refers to usage such as it was before the computer age. One begins to doubt whether those now employed in commerce still know how to add up columns of figures on paper, using the process known to accountants as 'casting'.

11. The present paragraph is based substantially on Lave 1988, chapter 8.

12. Lave's further statement (ibid.: 180) that 'this stands in opposition to the pervasive tendency in Western thought to dismiss the significance of active experience in the generation of cognitive processes' does not necessarily have to be accepted. For that matter this may also be a 'pervasive tendency' of Japanese thought, at least among educationalists.

13. The same idea can, apparently, be expressed in Japanese simply by *kujuku*, or '99' (Mori 1980: 474).

14. Both are instances of Turing's theorem, which establishes that 'to "process information" by computer is nothing more than to replace discrete symbols one at a time according to a finite set of rules' (Crump 1990: 150).

15. Kojima (1954: 60f) illustrates every step in the process of multiplying numbers of more than two digits.

16. I have witnessed a 12-year-old champion extract the cube-roots of fifteen 12-figure numbers in under a minute.

17. In the Kyoto Prefectural Commercial High School I have seen pupils move from an advanced abacus lesson in one classroom to instruction in the use of fifth-generation computers in an adjacent classroom.

18. The first character in this word means the brush used for writing: compare *hissha*, a 'writer', from the same basic root.

19. The normal term is *arabia sūji*, which corresponds to *rōmaji*, the term used for the Roman alphabet.

20. Arabic numerals were introduced into Europe at the beginning of the fourteenth century, but popular skills in elementary arithmetic only began to be common at the end of the eighteenth century. In Europe, however, the abacus still survives in Russia.

21. Reported in Shuzan Shunjū (1986: 62; 150–95).

22. There is neither need nor scope for any improvement. As a mechanical instrument it has, unlike a clock, neither motor nor hidden parts, and no accessories are ever needed.

23. *Jōjo* (multiplication and division) was dropped from the school curriculum in 1949 (Kubo 1986: 185).
24. The popularity of *soroban jukus* is on much the same level as that of those teaching piano or calligraphy (Kubo 1986: 153).
25. Kubo (1986: 117) even suggests that the abacus is taught in the *yōchien*, or Japanese kindergartens.
26. There appear to be no restrictions about running a *soroban juku*, and many are small private businesses, not affiliated to any national organisation. Such affiliation is however advantageous, not only for guaranteeing standards, but also for supplying teaching materials and organising competitions. Kubo (1986: 136f) gives a number of statistics relating to the use of the abacus.
27. Touch-typing is hardly possible with written Japanese. Until a few years ago every one of thousands of *kanji* had to be searched for on the matrix keyboard of the Japanese mechanical typewriter, a laborious process bearing no relation to touch-typing. Now the Japanese word-processor (*wāpuro*) has simplified the process by allowing all words to be typed in *rōmaji*, and displaying them in first instance in *hiragana* – one of the two *kana* syllabaries. The operator can then opt to convert selected words into *kanji* by using a special key. This process is also laborious, and in practice excludes touch-typing.
28. Kubo (1986: 113, 168) repeatedly stresses this point. Research carried out by Hatta (1986: 51f) suggests that the operation of the abacus by a proficient user is a function of the right hemisphere of the brain, and not the left as it is with a learner. When this stage is reached, the user is no longer consciously involved in the logical processes of arithmetic, any more than a virtuoso pianist is involved in problems of reading a score: in both cases, the processes are subconscious, and purely automatic, and the problems arising are at a quite different level of interpretation. The process of transforming the learner into a virtuoso unfolds on a sort of plateau of skill, from which only the top performers ever climb to a higher level. Perfect accuracy is an essential pre-condition for this transformation, which is then an inherent attribute of the proficient performer. The Japanese take for granted that mistakes are not made in the use of the abacus.
29. There is little experimental confirmation of this statement but see Yamada's work relating to alternative forms of keyboard input (Yamada 1983: 48f).
30. Sharp has even produced a combined abacus and *dentaku*.
31. Significantly the suffix '-en' (for Yen) is almost added by the teacher to numbers spoken out loud during a lesson.
32. This is one of the conclusions from the massive survey carried out by the National Abacus Education Union and reported in Shuzan Shunjū (1986: 8; 150–95).
33. *Buraku bokusu* is now a recognised neologism in Japan.
34. Stuller (1987: 66f) claims that the abacus is particularly well-suited to teaching arithmetic to handicapped or deprived children.
35. The book is a miscellany of short excerpts from newspaper and magazine articles, whose only common property is their unbridled enthusiasm for the abacus.

36. Kubo (1986: 118) gives ¥10,000 as the average cost of a month's lessons in a *soroban juku*: this is about $60.
37. Kubo (1986: 180) mentions in particular Nita and Yokota in Shimane prefecture, and Ono in Hyogo prefecture.
38. The fact that the abacus is displayed in a vertical position means that the beads must hold their position on the columns by friction. This is taken care of in the design.

## 10 GAMES

1. Lexical of analysis of *kakurembō*, the Japanese for 'hide and seek', suggests that the preferred venue is the village cemetery.
2. The Japanese *asobi* translates both 'play' and 'game', but it connotes the former rather more than the latter, for it hardly imports the idea of winning and losing. For this one can use *shiai* (in which the root *shi* means 'testing' and *ai* suggests 'meeting'), *shōbu* (in which the two components mean 'win–lose') or *kyōgi* (where *kyō* means 'compete' and *gi*, 'skill'). These three words are not quite synonymous, and their meaning depends on context. A game of tennis is *shōbu*, where a baseball game is *shiai*, leaving *shōbu* with the connotation of a fight to the finish (Whiting 1977: 124). *Kyōgi* is a definite contest, or sporting event.
3. Geertz's study of the Balinese cockfight is an exemplary proof of this statement (Geertz 1973, ch. 15).
4. *Go* is not strictly Japanese, since it was introduced into Japan from China in the eighth century. It brought with it a good deal of number symbolism, so that of the 361 points on the board, the one at the centre, known as the *taikyoku*, represents the primordial principle of the universe, whereas the remaining 360 represent the degrees of latitude. The 181 black, and the 180 white stones (*ishi*) with which the game is played, represent night and day, or yin and yang (Chamberlain 1974: 213f).
5. It is no coincidence that the same is true of computer programs: this is why it has been possible to develop such a wealth of computer games in the present generation.
6. In the present context this is hardly a pleonasm.
7. True in many cultures this statement, on the face of it, is somewhat reprehensible in its denial of the moral dimension expressed by such precepts as 'it's playing the game that counts, not winning'.
8. Poker addicts may well assert that bluff is not numerical, but in fact the top players do not resort to it (Alvarez 1984: 135; Crump 1990: 175). If you want to win at poker you play it as a numbers game pure and simple.
9. Cricket is perhaps a counter-example. A captain who declares an innings closed, so as to put the other side into bat, makes his decision on numerical factors, relating the number of runs scored to the time left for play.
10. To take the case of golf, the course has eighteen holes, each with its own measured length, and ranked in order of difficulty, which factor also determines the bogey (3, 4 or 5) rating of each hole. The latter add up to determine the number of strokes which is par for the course, on the basis of which each individual player calculates his own handicap, down to 24 below par. This in turn provides the basis for an ordered ranking of the

players in any actual game, as well as for determining the order of holes at which the weaker players will be allowed an extra stroke.

11. In the small Japanese magazine rack in the Okura Hotel in Amsterdam, three out of the periodicals offered for sale relate exclusively to golf. (This also says something about the up-market clientele of the hotel.)

12. The present analysis on the basis of these terms is derived from Caillois (1955).

13. 'Gambling', according to the *Concise Oxford Dictionary*, comes from the obsolete (sixteenth century) 'gamel', variant of Middle English 'gamen' (game). Japanese has a distinct word, *to* (generally *tobaku*), for 'gambling'. *Kake*, the corresponding *kun* reading of the same *kanji*, is the normal word for a 'bet' or 'wager'. See also note 2 above.

14. Such results can be produced mechanically (or electronically) by machines specially designed for this purpose: the obvious Japanese case is *pachinko*, a machine game for a single player, who must direct a constant stream of small metal balls into a winning slot.

15. The extent of this commitment is stupefying. Barthes records that in Japan the aggregate turnover of *pachinko* parlours exceeds that of department stores (Barthes 1970: 39).

16. *Ken* is the *on* reading of a *kanji* based on the *hand* radical. The normal word for 'first' is *kobushi*, which is the corresponding *kun* reading.

17. One such game, *tora* (tiger) *ken*, makes use of the whole body, in which case the two competing players must enter the scene, simultaneously, from outside (Linhart 1991).

18. These are *mue* (0), *ikkō* (1), *ryan* (2), *san* (3), *sū* (4), *gō* (5), *roma* (6), *chei* (7), *pama* (8), *kai* (9) and *tōrai* (10).

19. The *kanji* representation means literally 'wisteria eight', but the use of the *on* reading, *tō*, in place of the normal *fuji* for 'wisteria', suggests some sort of pun. On this point see Linhart (1991).

20. The whole history is given in Linhart (1991), which provides much of the historical material for the present section.

21. *Janken* was used to select the trainees from the first draft of volunteers for the *kamikaze* suicide bombers at the end of the Pacific War (Prof. T. Fuse, personal communication).

22. In the case observed (in March 1987), these were two boys, Yoshiki (11), Wataru (7) and a girl, Azusa (5), all children of Prof. Yoshio Yano of the Kyoto Education University.

23. Recorded at Gokayama (Toyama prefecture) in 1980.

24. On average a *decisive* position occurs every ninety-seven to ninety-eight rounds. In forty-nine rounds, therefore, there is a better than even chance of such a position occurring, just as there is a better than even chance of a six occurring with four throws of a dice.

25. The necessary mathematics belongs to a branch of statistics known as the theory of Markov chains.

26. In this case the mathematics is combinatorial, and Markov chains play no part.

27. This is certainly the conclusion of Hardy and Wright (1945: 116f). I also discuss *nim* in Crump (1990: 120).

28. *Go* is played on a flat board by the players alternately placing counters on the points of intersection of a $19 \times 19$ matrix. These are black for the

opening player, and white for his opponent. The object is to surround, and thereby eliminate, one's opponent's counters, so that, at the end of the game, the winner has placed the most counters. Iwamoto (1976) gives a short introduction to the history, rules, elementary strategy and the system for ranking the players.

29. This is the conclusion reached by Unger (1987: 113), citing Dreyfus (1979: 291f), but then the games cited in support are *nim*, and the even more elementary *tic-tac-toe* (or in England, *noughts and crosses*).

30. Kawabata (1972) gives a fictional account of what this involved in the world of *go*.

31. See chapter 9, note 28.

32. The *go* player has in principle much more time to think. In practice play alternates between a rapid succession of moves by both players, ending in a position where one player or the other must spend several minutes thinking out his future strategy.

33. Mahjong, although of Chinese origin, is still extremely popular in Japan.

34. Leggett (1966) gives both the rules of *shōgi* and elementary strategy. The game has so much in common with chess that the two must have some common origin. The *shōgi* board has, however, nine squares each way, and what is more, a captured enemy piece can be held in reserve by the captor, to be held in reserve as his piece, to be placed anywhere on the board instead of a normal move. This possibility, not found in any other form of chess, gives *shōgi* a quite distinctive form. Chamberlain (1974: 90) also states that *shōgi* was originally introduced from China, centuries ago.

35. This is simply the Japanese equivalent of the Chinese *tao*. Note, however, its occurrence in *shintō*, 'the way of the gods' and *bushidō*, the way of the *bushi*, that is, the *samurai*.

36. If *kadō* is the 'flower way', its followers are engaged in the more familiar *ikebana*, literally 'arranging flowers'. *Kadō*, written with another initial *kanji*, is also the 'song way', whose followers write the 31-syllable poems known as *tanka*, literally 'short songs'.

37. Common to all these schools is the teaching of the principles of *seishin*, best translated as 'spirit' in a positive moral sense, as in '*team* spirit'. On this point, according to Smith, 'the authority of the head of the school is absolute, far more so than that exercised by the heads of "real" households. It is believed that the more demanding the teacher, the more valuable the training, and it is at the head of some of these groups that Japan's true autocrats are to be found' (Smith 1983: 99).

38. The numerical basis is reflected in the word *kyūsū* for a mathematical series of progression.

39. History records one case of a pure competitive ranking quite outside the *kyū-dan* system. This is the archery contest, known as the 'Tōshiya', which first took place in the year 1606 in the west porch of the Kyoto temple of Sanjūsangendō described in chapter 8. A challenger to the title-holder, within a period of twenty-four hours beginning at 6 p.m., had to shoot as many arrows as possible from the south end of the porch that would reach the north end. This was a considerable feat, because the porch was long, and its roof, low. In the year 1686, the final challenger, the 22-year-old Wasa Daihachirō, beat the title-holder Hoshino, by

shooting 13,053 arrows, of which 8,133 reached the far end of the porch. Since then the contest has never been held.

40. The normal word for competition, *kyōsō*, is a neologism coined by Yukichi Fukuzawa in the late nineteenth century for use in the translation of economic texts in English. Although the word is now equally used in the context of sport and games, the connotations of its two component *kanji* are negative, as if to suggest that in an ideal world *kyōsō* would play no part. This may still be true of the ideology of the various *dō*, but if so, it has been completely corrupted by the competitive ethic of sport in the present century. Another word, *shiai*, has more positive connotations, in that it implies a 'coming together' in order to be 'tested', as in a competitive examination, but a Japanese would not regard *kyōsō* and *shiai* as synonyms.

41. The same principle governs the Elo ratings in international chess.

42. For *go*, the consequences of this for the top hierarchy are extremely evocatively portrayed in Kawabata (1972).

43. *Kendō*, as a *dō*, is a twentieth century development from traditional *kenjutsu* or 'sword art', which became a sport, rather than a fight, after bamboo swords (*shinai*) and protective gear were introduced in the eighteenth century. At the end of the Pacific War the occupying authority banned *kendō* as furthering militarism, but it was revived shortly after the end of the occupation.

44. For *aikidō*, Westbrook and Ratti (1970: 30f) is an English text explaining this in some detail.

45. The material relating to *sumō* comes from Reader (1989), while figure 10.1 is based on Simmons (1986).

46. These occur in the odd-numbered months, and alternate between Tokyo (in January, May and September), and one of three other centres, Osaka (in March), Nagoya (in July) and Fukuoka (in November).

47. There does not seem to be any particular significance in this number, and it may in special circumstances be exceeded.

## 11 THE ECOLOGY OF NUMBERS

1. The date given below, which is generally accepted by modern historians, occurs in the year AD 712.

2. As I noted in chapter 7, this is still the official method of dating documents. The Wadō era coincided exactly with the reign of the Emperor Gemmei (707–15). Most of the dates in the *Kojiki*, particularly of the deaths of previous emperors, are given according to the *kanshi* sixty-year cycle.

3. A hereditary title established as part of the Temmu reform of the late seventh century.

4. The old Japanese binary numeral system is given in Miller (1967: 337, table 9).

5. A good short description of what this meant for Japan is given in the Introduction to Philippi (1968).

6. The extreme case is that of the Balinese, who work with an incredibly intricate permutational calendar (Geertz 1973: 392f). The Hindu temple

rituals which it regulates are closely tied to the local cycle of wet-rice cultivation.

7. I describe this in relation to Japan in Crump (1986).

8. The most interesting issue here is the status and content of the institutional rules governing the market. The *rationality* of dealing can be profoundly altered by ignoring these rules, which is what the Recruit Cosmos scandal in Japan was all about. In 1989 it led to the resignation of a prime minister, Takeshita, and to the near defeat of the ruling Liberal Democratic Party.

9. The name means 'temple township street', and there are still one or two temples on it, offering perhaps the chance of consulting an *omikuji* oracle. The majority, however, were suppressed during the Edo period, when the shoguns were intent on limiting the power of the Buddhist priests.

10. It is of interest to note that Buddhist temples, in their choice of names (*kaimyō* or *hōmyō*) to be bestowed on the recently deceased, make no use whatever of *seimeigaku.*

11. Japanese newspapers regularly carry advertisements for the sale of cars with auspicious number plates, generally characterised by symmetry in the numbers, the repeated occurrence of 8, and the absence of 4.

12. One would suppose that there was somewhere a 'VAT 68', but just try asking for it at your local liquor store.

13. Newsweek (July 1, 1991) reports the name of the assistant bishop of Paris as André Vingt-trois (=23); compare chapter 5, note 37.

14. A year or two back, a proof claimed for Fermat's last theorem (dating from the seventeenth century, but so far never proved), showed at least the will to tackle the most difficult problems. It is no objection that the attempted proof relied upon an extended computer program, for this was also the approach adopted for the successful proof (by American mathematicians) of the equally intractable *four colour theorem.* It is, however, characteristically Japanese to adopt the programmatic methods inherent in the use of computers for solving even the deepest problems of pure mathematics.

# References

Ahern, E.M. 1981. *Chinese Ritual and Politics.* Cambridge University Press.

Alvarez, A. 1984. *The Biggest Game in Town.* London, Fontana.

Ayer, A.J. 1963. *The Concept of a Person and Other Essays.* London, Macmillan.

Barthes, R. 1967. *Elements of Sociology.* London, Jonathan Cape.

Barthes, R. 1970. *L'Empire des Signes.* Geneva, Skira.

Bellah, R. 1985. *Tokugawa Religion.* New York, Free Press.

Berque, A. 1982. *Vivre l'espace au Japon.* Paris, P.U.F.

Berque, A. 1986. 'The sense of nature and its relation to space in Japan'. In J. Hendry and J. Webber (eds) *Interpreting Japanese Society.* Oxford, JASO, pp. 100–10.

Berque, A. 1987 (ed.). *La Qualité de la ville. Urbanité française, urbanité nippone.* Tokyo, Maison franco-japonaise.

Blacker, C. 1986. *The Catalpa Bow: A Study of Shamanistic Practices in Japan.* London, Allen & Unwin.

Blofeld, J. 1976. *I Ching: The Book of Change.* London, Allen & Unwin.

Bolter, J.D. 1986. *Turing's Man.* Harmondsworth, Penguin.

Brainerd, C.J. 1973. 'Mathematical and behavioral foundations of number'. *Journal of General Psychology* 88: 221–81.

Bray, F. 1986. *The Rice Economies: Technology and Development in Asian Societies.* Oxford, Basil Blackwell.

Caillet, L. 1986. 'Time in the Japanese ritual year'. In J. Hendry and J. Webber (eds) *Interpreting Japanese Society.* Oxford, JASO.

Caillois, R. 1955. 'Structure et classification des jeux'. *Diogène* 12: 72–88.

Carroll, L. 1972. *Kagami no kuni no Arisu (Alice through the Looking Glass,* trans. K. Shono). Tokyo, Fukuonkan Shoten.

Chamberlain, B.H. 1974. *Japanese Things.* Tokyo, Charles E. Tuttle.

Crump, T. 1978. 'Money and number: the Trojan horse of language'. *Man,* n.s., vol. 13, pp. 503–18.

Crump, T. 1980. 'Trees and stars: Graph Theory in southern Mexico'. In J.C. Mitchell (ed.) *Numerical Techniques in Social Anthropology.* Philadelphia, ISHI, pp. 163–90.

Crump, T. 1986. 'The Pythagorean view of time and space in Japan'. In J. Hendry and J. Webber (eds) *Interpreting Japanese Society.* Oxford, JASO.

Crump, T. 1988. 'Japanese as a private language'. Paper presented to Third Congress of the European Association of Japanese Studies, Durham.

Crump, T. 1989. *The Death of an Emperor.* London, Constable.
Crump, T. 1990. *The Anthropology of Numbers.* Cambridge University Press.
*Daihyakka Jiten* (The Great Encyclopedia) 1951. Tokyo, Heibonsha.
Dale, P.N. 1986. *The Myth of Japanese Uniqueness.* London, Routledge.
Davis, P.J. and Hersh, R. 1983. *The Mathematical Experience.* London, Pelican.
de Francis, J. 1984. *The Chinese Language: Fact and Fantasy.* University of Hawaii Press.
de Saussure, F. 1967. *Cours de linguistique générale.* Paris, Payot.
Dixon, R.M.W. 1980. *The Languages of Australia.* Cambridge University Press.
Dreyfus, H.L. 1979. *What Computers Can't Do.* New York, Harper.
Dreyfus, H.L. and Dreyfus, S.E. 1986. *Mind over Machine: The Power of Human Intuition and Expertise in the Era of the Computer.* New York, Free Press.
Durkheim, E. 1915. *The Elementary Forms of the Religious Life.* London, George Allen & Unwin.
Embree, J.F. 1939. *Suye Mura: A Japanese Village.* University of Chicago Press.
Fujimoto, T. and Sakai, H. 1968. *Ronri tetsugaku ronkō* (translation of L. Wittgenstein's *Philosophical Investigations*). Tokyo, Hosei University Press.
Fukuzawa, Y. 1966. *Autobiography.* New York, Columbia University Press.
Gardner, M. 1970 (ed.), *The Annotated Alice.* Lewis Carroll. London, Pelican.
Geertz, C. 1973. *The Interpretation of Cultures.* New York, Basic Books.
Gellner, E. 1972. 'The savage and the modern mind'. In R. Horton and R. Finnegan (eds) *Modes of Thought.* London, Faber & Faber.
Gerschel, L. 1962. 'La conquête du nombre: des modalités du compte aux structures de la pensée'. *Annales ESC* 17: 691– 794.
Goodman, G.K. 1986. *Japan: The Dutch Experience.* London, Athlone Press.
Gregory, R.L. 1969. 'On how little information controls so much behaviour'. In C.H. Waddington (ed.) *Towards a Theoretical Biology*, vol. 1. Chicago, Aldine Publishing Company.
Grew, J.C. 1945. *Ten Years in Japan.* New York, Simon & Schuster.
Gudeman, S. 1986. *Economics as Culture.* London, Routledge & Kegan Paul.
Guillain, R. 1979. *La Guerre au Japon.* Paris, Seuil.
Hallpike, C.R. 1979. *The Foundations of Primitive Thought.* Oxford, Clarendon Press.
Hardy, G.H. 1940. *A Mathematician's Apology.* Cambridge University Press.
Hardy, G.H. and Wright, E.M. 1945. *The Theory of Numbers.* Oxford University Press.
Hatta, T. 1986. *Nobiru sodatsu kodomo no nō* (The Brain of the Growing Child) Tokyo, Rodo Keizaisha.
Hobsbawm, E. and Ranger, T. (eds) 1983. *The Invention of Tradition.* Cambridge University Press.
Horn, W. 1975. 'On the selective use of sacred numbers and the creation in Carolingian architecture of a new aesthetic based on modular concepts'. *Viator* 6: 351–90.

198    *References*

Hurford, J.R. 1987. *Language and Number.* Oxford, Basil Blackwell.
Itoh, T. 1972. *Traditional Domestic Architecture of Japan.* New York, Weatherhill.
Iwamoto, K. 1976. *Go for Beginners.* London, Penguin.
Janelli, D.Y. 1977. *Logical contradictions in Korean learned fortune-telling.* University of Pennsylvania, Ph.D. thesis.
Johnston, R.F. 1985. *Twilight in the Forbidden City.* Oxford University Press.
Junod, H.A. 1927. *The Life of South African Tribe.* London, Macmillan.
*Kanyo Kotowaza Jiten* (Colloquial Dictionary of Proverbs) 1988. Tokyo, Shogakkan.
Kawabata, Y. 1972. *The Master of Go.* Tokyo, Charles E. Tuttle.
Keesing, R.M. 1981. *Cultural Anthropology: A Contemporary Perspective.* New York, Holt, Rinehart & Winston.
Kobayashi, M. 1951. *Seimei no hikari* (The Light of Full Names). Tokyo, Kaiun Shinbunsha.
Kojima, T. 1954. *The Japanese Abacus: Its Use and Theory.* Tokyo, Charles E. Tuttle.
Kubo, J. 1986. *Soroban no miryoku o saguru* (Exploring the Charm of the Abacus). Tokyo, Akatsuki.
Lancy, D.F. 1983. *Cross-cultural Studies in Cognition and Mathematics.* New York, Academic Press.
Lave, J. 1988. *Cognition in Practice.* Cambridge University Press.
Leach, E.R. 1966. *Rethinking Anthropology.* London, Athlone Press.
Leggett, T. 1966. *Shogi: Japan's Game of Strategy.* Tokyo, Charles E. Tuttle.
Lévi-Strauss, C. 1985. *La Potière Jalouse.* Paris, Plon.
Lewis, D.C. 1986. 'Years of calamity': *yakudoshi* observances in a city'. In J. Hendry and J. Webber (eds) *Interpreting Japanese Society.* Oxford, JASO.
Linhart, S., 1991. 'Ritual in the Ken game'. Paper given at Japanese Anthropology Conference, Leiden, 1990.
Lodge, D. 1989. *Nice Work.* Harmondsworth, Penguin.
MacFarland, H.N. 1967. *The Twilight of the Gods.* New York, Macmillan Company.
Masai Y. 1987. 'Formes et fonctions des métropoles japonaises'. In A. Berque (ed.), *La Qualité de la Ville. Urbanité française, urbanité nippon.* Tokyo, Maison franco-japonaise, pp. 66–80.
Menninger, K. 1977. *Number Words and Number Symbols.* Cambridge, Mass., MIT Press.
Miller, G.A. 1967. *The Japanese Language.* University of Chicago Press.
Miller, G.A. 1971. *Japanese and Other Altaic Languages.* University of Chicago Press.
Miller, G.A. 1982. *Japan's Modern Myth.* New York, Weatherhill.
Miller, G.A. 1986. *Nihongo: in Defence of Japanese.* London, Athlone Press.
Miyako, J. 1980. *Seikatsu no naka no shūkyō.* (Religion in Life). Tokyo, NHK Books.
Miyazaki, N. 1985. *Shikoku Henro: Rekishi to Kokoro* (Shikoku Pilgrims: History and Motive). Osaka, Toki Shoten.
Moeran, B. 1985. 'Confucian confusion: the good, the bad and the noodle western'. In D. Parkin (ed.) *The Anthropology of Evil.* Oxford, Basil Blackwell, pp. 92–109.

Mori, M. 1980. *Meisū Sūshi Jiten* (A dictionary of number compounds). Tokyo, Iwanami.

Murō, T. 1983. *Gendai no Shūkyō: Risshō Kōseikai* (Risshō Kōseikai: a present day religion). Tokyo, Sōshuppan.

Murray, A. 1978. *Reason and Society in the Middle Ages.* Oxford, Clarendon Press.

Nakayama, S. 1969. *A History of Japanese Astronomy.* Cambridge, Mass., Harvard University Press.

Nakayama, S., Swain, D.L. and Yagi, E. 1974 (eds). *Science and Society in Modern Japan.* Cambridge, Mass., MIT Press.

Needham, J. 1969. *With the Four Seas.* London, Allen & Unwin.

Needham, J. 1980. *Science and Civilisation in China,* vol. 5, part 4. Cambridge University Press.

Nelson, A.M. 1974. *The Modern Reader's Japanese–English Character Dictionary.* Tokyo, Charles E. Tuttle.

Norbeck, E. 1955. 'Yakudoshi: a Japanese complex of supernatural beliefs'. *Southwestern Journal of Anthropology* XI: 105–20.

Ogawa, D.M. and Grant, G. 1990. *Hawaii's Yakudoshi Guidebook.* Honolulu, Nippon Golden Network.

Ohnuki-Tierney, E. 1984. *Illness and Culture in Contemporary Japan.* Cambridge University Press.

Ohnuki-Tierney, E. 1989. *The Monkey as Mirror: Symbolic Transformations in Japanese History and Ritual.* Princeton University Press.

Okubo, T. 1978. *Ichiokunin no kokugo kokuji mondai* (The Language Problem of 100,000,000 People). Tokyo, Sanseido.

Okuya, T. 1982. *Seimei handan nyūmon* (Introduction to the Interpretation of Full Names). Tokyo, Shinsei Shuppansha.

O'Neill, P.G. 1972. *Japanese Names.* New York and Tokyo, Weatherhill.

O'Neill, P.G. 1984. *A Reader of Handwritten Japanese.* Tokyo, Kodansha International.

Oshima, T. 1959. 'Shinkō to nenjū gyōji' (Beliefs and annual observances). *Nihon Minzokugaku Taikei* 7: 67–116.

Oya, S. 1980. *Wasan izen* (Former Japanese Calculation). Tokyo, Chūō Koronsha.

Palmer, M. 1986. *T'ung shu: The Ancient Chinese Almanac.* London, Rider & Co.

Philippi, D.L. 1968. *Kojiki* (translation). University of Tokyo Press.

Piaget, J. 1952. *The Child's Conception of Number.* London, Routledge & Kegan Paul.

Picken, S.D.P. 1980. *Shinto: Japan's Spiritual Roots.* Tokyo, Kodansha International.

Pye, M. 1977. *Everyday Japanese Characters.* Tokyo, Hokuseido Press.

Reader, I. 1985. 'The Thousand Day Pilgrimage of Mount Hiei'. *Kansai Time Out,* March.

Reader, I. 1987a. 'Shichi-Fuku-Jin: the Seven Gods of Good Fortune'. *Kansai Time Out,* January.

Reader, I. 1987b. 'From asceticism to the package tour'. *Religion* 17: 133–48.

Reader, I. 1988. 'Miniaturization and proliferation: a study of small scale pilgrimages in Japan'. *Studies in Central and East-Asian Religions* I.1: 50–66.

Reader, I. 1989. 'Sumo: the recent history of an ethical model for Japanese society'. *International Journal of the History of Sport* 6.3: 285–98.

Ronan, C.A. and Needham, J. 1978. *The Shorter Science and Civilisation in China: 1.* Cambridge University Press.

Ronan, C.A. and Needham, J. 1981. *The Shorter Science and Civilisation in China: 2.* Cambridge University Press.

Russell, B. 1920. *An Introduction to Mathematical Philosophy.* London, Allen & Unwin.

Shimano, J. 1956. *Oriental Fortune Telling.* Tokyo, Charles E. Tuttle.

Shimodaira, K. 1986. *Nihonjin no sūgakukanki* (The Japanese Feeling for Mathematics). Kyoto, PHP Kenkyujo.

Shorto, H.L. 1963. 'The 32 myos in the medieval Mon kingdom'. *Bulletin of the School of Oriental and African Studies* xxvi: 572–91.

Shorto, H.L. 1967. 'The Dewatau Sotopan: a Mon prototype of the 37 Nats'. *Bulletin of the School of Oriental and African Studies* xxx: 127–41.

Simmons, D. 1986. *The Kanroku Scale.* Kansai Time Out, January, p. 30.

Smith, D.E. 1958. *History of Mathematics*, vol. I. New York, Dover Publications.

Smith, R.J. 1974. *Ancestor Worship in Contemporary Japan.* Stanford University Press.

Smith, R.J. 1983. *Japanese Society: Tradition, Self and the Social Order.* Cambridge University Press.

Smith, T.C. 1977. *Nakahara: Family Farming and Population in a Japanese Village, 1717–1830.* Stanford University Press.

Snodgrass, A. 1985. *The Symbolism of the Stupa.* Ithaca, N.Y., Cornell University Press.

Sontag, S. 1989. *Traditions of the New* (Huizinga Lecture for 1989). Leiden, N.R.C. Handelsblad.

Stuller, J. 1987. 'New conquests for the ancient abacus'. *Reader's Digest.* March, pp. 65–9.

Sugimoto, S. and Swain, D.L. 1978. *Science and Culture in Traditional Japan: AD 600–1854.* Cambridge, Mass., MIT Press.

Tambiah, S.J. 1976. *World Conqueror and World Renouncer.* Cambridge University Press.

Tanaka, N. 1988. *Taishō Tennō no Taisō* (The Taishō Emperor's Funeral). Tokyo, Daisan Shokan.

*The Threefold Lotus Sutra* 1975. New York and Tokyo, Weatherhill and Kosei.

Toya, S. 1982. *Nihon shuzanshi* (History of Abacus Calculation in Japan). Tokyo, Akatsuki.

Toyota, A. 1984. *Yoku tsukuwareru shinbun no kanji to jukugo* (The Correct Use of Newspaper *Kanji* and Compounds). Tokyo, Bonjin.

Tsuboi, H. 1970. 'Nihonjin no seishikan' (The Japanese way of life and death). In *Minzokugaku kara mita Nihon* (Japan in the Light of Folklore). Tokyo, Kawade Shobōsha, pp. 18–34.

Tuchman, B. 1984. *The March of Folly: From Troy to Vietnam.* Glasgow, Collins.

Turner, V. 1974. *Dramas, Fields and Metaphors.* Ithaca, NY, Cornell University Press.

Unger, J.M. 1987. *The Fifth Generation Fallacy.* New York, Oxford University Press.

Valentine, J. 1986. 'Dance, space, time and organization: aspects of Japanese cultural performance'. In J. Hendry and J. Webber (eds) *Interpreting Japanese Society*. Oxford, JASO.

Vygotsky, L.S. 1962. *Thought and Language*. Cambridge, Mass., MIT Press.

Westbrook, A. and Ratti, O. 1970. *Aikido and the Dynamic Sphere*. Tokyo, Charles E. Tuttle.

Whiting, R. 1977. *The Chrysanthemum and the Bat: The Game Japanese Play*. Tokyo, Permanent Press.

Williams, B. 1978. *Descartes: The Project of Pure Enquiry*. London, Penguin.

Wittgenstein, L. 1958. *Philosophical Investigations*. Oxford, Basil Blackwell.

Yamada, H. 1983. *Certain Problems Associated with the Design for Input Keyboards for Japanese Writing*. Tokyo, University Department of Information Science.

Yano, Y. 1973. 'Kazoku no komyunikēshon seikatsu' (Communication life of the family). In T. Ota, Y. Fujioka, T. Nogawa and T. Inoue (eds) *Toshi ni okeru kazoku no seikatsu* (The Life of the Family in the Year). Kyoto University Institute for Humanistic Studies.

Yoneyama, T. 1974. *Gion Matsuri* (The Gion Festival). Tokyo, Chūō Koronsha.

Yoneyama, T. 1990. *Nihonjin koto hajime monogatari* (In the beginning: training out the roots of ideas and customs we Japanese take for granted). Tokyo, PHP Kenkyūjo.

Yoshida, M. 1987. *Nihon kagakushi* (History of Japanese Science). Tokyo, Kodansha.

Yoshino, H. 1983. *Inyō gogyō to Nihon no minzoku* (Yin-yang, the five elements and Japanese folk customs). Tokyo, Jimmon Shoin.

Zengoku Chōsa. 'Sansū ni kansuru (chūtoshite chusan) ishiki chōsa. Kōritsu shōgakkō no kyōshi, jidō, hogosha ni taishite' (A cognitive investigation into arithmetic (principally abacus calculation) in relation to children, teachers and parents of state primary schools). *Shuzan shunjū* 62: 150–95.

# Index

abacus 3, 8, 20, 126–38, 159;
accuracy 134; Chinese 126f, 135;
Japanese 127f; speed 134; *see also*
addition; division; *jōjo*; *juku*
(*soroban*); *kagen*; markets;
multiplication; subtraction;
*shuzan*
acorn 50
acquired (characteristics) 51
active 51
addition 11, 17, 33; abacus 127f; *see
also kagen*
affinity (between man and woman)
84f
age chart *see nenrei hayamihyō*
*agon* 141f
agriculture 3, 107f, 156f
*aikidō* 150f
*aiko see ken*
Ainu 23
*aishō* 72f
*alea* 141f, 149
algebra 11, 46; of metonymy 48, 145;
*see also* equations
algorithm 17, 37; abacus 132
Alice 34, 161
almanac *see koyomi*; *kyūsei*; *T'ung
Shu*
Altaic 23
alternation 97f, 104
*Ama-no-hashidate* 51
Ameterasu Omikami 107
analytical geometry 124
ancestor 106f; *see also ihai*
Angkor Wat 38, 161
antagonism 39–41

*antei see* stable
*anzan* 134
apples 19–20
Arabic numerals 17–18, 35–6; *see
also sanjō sūji*
archery 150
area *see* measurement
arithmetic 1, 7, 36, 56; *see also
anzan*; numbers (operational use)
Asia, South and East 157
astrology 161; *see also ekikyō*
astronomy 5, 112f; western 5
August 106
autumn 49

backgammon 147
Bali *see* cockfight
banks 134
*banshō* 35
*banzuke see sumō*
baseball 141f, 146, 160
beads *see* abacus; *shuzan*
*beiju* 61
binary *see* numbers
birth 100, 105; grain spirits 107f; *see
also chūin*; *hatsu-miyamairi*;
*hyakkanichi*; *shonanuka*
*black box* 136
blue 59
*bōken* 13
Bon (Festival) 100, 106f
Book of Changes *see ekikyō*; *I Ching*
Borobudur 38, 161
boules 139
bowls 139
*Brahma*, egg of 125

*bricolage* 157
bridge 149
Buddhism 52, 100, 104f, 125, 156f;
    Mahayana 59; Nichiren 73; and
    Shinto 104f, 157; *see also*
    *butsudan*; death (rites); *ihai*;
    *Kannon*; *Kegon*; *Risshō*
    *Kōseikai*; *Shingon*
*butsudan* 74

calculation 7; written 132, 135; *see*
    *also sanjō sūji*
calendar 5, 52, 108, 113; Gregorian
    108f
calligraphy 136, 150
Cambodia 3
car-racing 140
cardinal *see* number
cards: playing 34
Carroll, Lewis 34
Cartesian co-ordinates 124
centipede 56
century 19
*chadō see* tea ceremony
chance *see alea*
charm *see fudu*
chaos 34–5
chess 147, 153
*chi* 66
Chiba 66
*chikatetsu* 114
children 67, 101, 139, 145
China 14, 16–17, 25, 55, 101, 109,
    147, 155; *see also* Mandarin;
    numerals
Chinese characters *see kanji*
*chiten* 114
Chiyoda 65f
*chōka* 43
Christianity 47, 61, 108f
*chūin* 100, 104
Chūō 65
Circle Line 65; *see also kanjōsen*
cockfight 142
Cockney 54
cognition 2, 56, 130; and computers
    149
colours 59
*communitas* 119
computers 18, 134, 136f; and

cognition 149; *dentaku* 136, 160;
    fifth generation 138; *pasocon* 158
Confucianism 47
constellations 84
cosmology 5
cosmos 4, 34–5, 53
counters *see josūshi*
cube roots 132
culture 2, 12, 130
cure 60
cycles 98, 104, 112; ritual 105f; *see*
    *also* agriculture; death; *kanshi*;
    life

*daikichi* 72
*daimyō* 67
*dan* (step) 116f, 136, 150–4
*dandambatake* 121
danger 13
darkness 35
day[s] 6, 60, 99, 112; in Japanese 24
death (rites) 60, 105f, 156; names
    75; *see also ihai*
decade 6
decimals 58
*dentaku see* computer
Descartes 124
*Deshima* 12
dictionary 31, 70
Diet 110
direction board *see hōiban*
dimension 52, 114
divination 76–80
division 17, 33
*dō* 150
dogs 58
*dōkon* 35
Durkheim 2–3
Dutch: proverb 46
dynamic 114

earth 37, 39, 67; *see also kanshi*
East 64
egg *see Brahma*
eight 54
eighty-eight 61f, 102, 117f
*ekikyō* 37, 54, 76, 89–94, 158–61
*ekisha* 68, 90
electronics 12
eleven 120f

Elo match-points 153
Emperor 66; *see also* Meiji
English 18, 48f; proverbs 46f
equations: algebraic 12; quadratic 12
equinox 5

fate *see yakudoshi*
February 21, 101, 111
feeling for figures *see sūjikankaku*
female 35
*fen* 28, 98
feudal 118
fifty 67
fingers *see janken*; *ken*
fire 4, 37, 39, 67
fish-market 159
five 47, 53; elements (*gogyō*) 38–43,
    67f
flood 4
floor-plan 122f
flower arrangement 136, 150
form *see katachi*; *mukei*; *yūkei*
fortune-telling 36, 76–95
forty-two 120f
four 53, 60; *see also* seasons
fractions 11, 28–9
frame 61; *see also* language
French 15, 18
*fuantei see* unstable
*fuda* 102
Fuji 116f, 150
Fūjiyama 124
*fukuon* 72
Fukuzawa 99
Futako 65

games 12, 34, 36, 139–54; rules 140
    (constitutive 140f, regulative 140,
    142); strategy 147–9; *see also*
    *agon*; *alea*; *dan*; *kyū*; ranking;
    schools
go 12, 140, 147f, 159; *see also*
    Kawabata
*gogyō* 38–41, 43, 70, 73, 81f
*gojūon* 29
gold 58
golf 139f, 146
*gōme* 116, 150
Goshirakawa (Emperor) 9

Gotanda 65, 161
greengrocer 24, 56
Gumma (prefecture) 107

*hachi jūhakkasho* 117f
Hades 54
Haguro 116
*haiku* 43–5
*hakke* 36–7, 85, 91f
*hakuju* 62
*hanhan* 89
*hankichi-hankyō* 72
*hanshi* 151
ha'porth 61
harmony 39–41
*hasami* 72
Hatsukaishi 66
*hatsu-miyamairi* 100
heaven 37; *see also kanji*
*hegira see* Islam
*Heisei* 21, 110
hexagrams 36, 85, 91f
*heya see sumō*
hide-and-seek 139
Higashiyama 125
*hinoe uma* 85
*hiragana* 27–8
Hirohito (Emperor) 110
Hiroshima 63, 66; *see also*
    Hatsukaishi; Itsukaishi
*hissan see* calculation (written)
*hōiban* 85f
*hōin see* seal
Hokkaido 64, 124
Hokuriku 64
Holland *see* Dutch: proverb; *rangaku*
home-town 124; *see also kokoro no*
    *furusato*
*honji-suijaku* 157
*honken* 144
Honshu 64
horoscope *see kyūsei*
horse: racing 140; *see also hinoe uma*
*hotoke* 106
hour 60, 98, 112
household *see* ancestors; *ie*; *ihai*
Humpty Dumpty 34, 36
hundred 56
hundred million 57
*hyakkanichi* 100

'I' 74
*I Ching* 37, 54, 76, 85
*ichiritsu* 72
*ie* 74, 106
*iemoto see sumō*
*ihai* 74, 106
Ikago 67
*imiake* 104
inborn (characteristics) 51
inch 49f
Indo-European 15, 23
infinite series 33, 35
infinity 57
insurance 13, 134, 158
integers 11
investment 158
*in-yō* 34–8; *see also yin–yang*
*i-ro-ha* 29
Ise 107
Islam 109
Itsukaishi 66

Jack of all trades 56
*janken* 6, 12, 143–7, 161
Japan Sea 66
Japanese *see* numbers
Java 38
*jibun* 98
jidai 97, 114
*jikan* 97
*jikoku* 97
Jimmu (Emperor) 109
*jin'i* 72
*jiten* 97, 114
-jō 125
*jōjo* 131f
*josūshi* 20–1, 24
*jūdō* 151
*jūjika* 61
*juku* 133; *kyōiku* 133; *soroban* 133f,
    137
Jupiter 39, 53
*jūryō see sumō*
Juso 66

kadō *see* flower arrangement
*kagen* 127
*kakuritsu* 13
*kami* 80, 106, 116; *see also ujigami*
Kamigyō 125

Kamo Mioya (shrine) 107
*kan* 39–41
*kana* 27; order of 29–30; *see also*
    *hiragana*
*kanji* 1, 7–8, 23, 61, 63, 70, 81, 155f;
    dictionary 30–1; *jōyō* 30, 156;
    order of 30–2; radicals 31; time
    97–9; *tōyō* 30; *see also hiragana*;
    *kana*; *seimeigaku*; numerals
kanjōsen 115f
*Kannon Buddha* 11, 117, 119f
*kanreki* 102
*kanroku* (scale) *see sumō*
Kansai 64
*kanshi* 9, 41–3, 81f, 111f
Kantō 64
Kasuga (shrine) 107
*katachi* (form) 124
*katayori* 72
*kawa* 67
Kawabata 148
Kawaguchi 67
*kazuken* 144
Kazuraki (Mount) 49
*Kegon* (Buddhism) 59f
*kekkon see* marriage
*ken* 143; *aiko* (draw) 143
*kendō* 149f
*ki* 67
*ki-* (relating to time) 98
*kichi* 77f, 87
*kigensetsu* 111
Kii (peninsular) 125
*kiju* 61
*kiken* 13
*kikkyō* 89
Kimura 67
*kishō* 81
Kita 125
Kitamura 67, 72
Kiyomizu (Temple) 122
Kōan (Emperor) 111
Kōbō Daishi 117
*kojiki* 155
*kōki* 109
*kokoro no furusato* 124
*kokuritsu* 153
Kōmei (Emperor) 110
*komusubi see sumō*
Korea 156

Korean 23
Kōtoku (Emperor) 110
*koyomi* 80f
Kudanshita 65
*kuji*: meaning of 80; *see also omikuji*
*kujihiki* 77
Kujūkuri 66
*kun see* numerals: *kanji*
*kuni* 64
*kyōshi* 151
Kyoto 10–11, 64, 102, 107, 115,
    125, 133, 145, 158
*kyō* 70f, 76f, 87
kyū 84, 150–4
*kyū-dan* 153; *see also* ranking
*kyūdō see* archery
*kyūsei* 76, 80–9
Kyushu 63f

lake 37
language: *meta-* 9; and
    neurophysiology 14; and numbers
    8, 14–32; (cyclic 16, 127; frame
    15, 17, 127); written 14; *see also*
    French; Indo-European;
    morpheme; Yoruba
leaf 49
left 125
life 105
light 35
line 114f
lines (in *I Ching*) 92f
logic 158
lottery 45; *see also omikuji*
*Lotus Sutra* 11, 119f
lower 125

*maegashira see sumō*
mahjong 147, 159
Mahomet 109
*maku no uchi see sumō*
male 35
*mamori* 102
*mandala* 10, 37–8, 85, 117, 125
Mandarin 25, 28
markets 118, 132
marriage 72, 105
Mars 39, 53
*Master of Go see* Kawabata
mathematics: actuarial 158, pure 1

Matsushima 51
May 24
measurement 18–29, 34
medicine 58f
Meiji (Emperor) 43, 59, 67, 110, 145
*meishi* 73f
memorial tablets *see ihai*
merchants 67
Mercury 39, 53
metal 39, 67
metaphor 94f, 144
metonymy 46, 48, 95, 116, 144
Mikuni 64
Minami 125
*minka* 122
minute 98
mirror (sacred) 107
Mitaka 65
Miyajima 51
*mizu* 67
module 122
Mommu (Emperor) 110
Momoyama 145
money 28
month 4, 6, 112
moon 4, 53, 99, 101, 112
morpheme 16–17
Motoori (Norinaga) 45
mountain 35, 37, 116f
movement 114
*mukei* 124
multiplication 11, 17, 33; abacus
    129f (*see also jōjo*)
*mura* 67
music 12

Nachi 117
Nagasaki 12
Nakagyō 125
Nakasone 146
names: change of 75; numbers in 9,
    63–75; *see also* death; *meishi*;
    *seimeigaku*
Nanako 67
Nara 49, 107
National Abacus Education
    Association 132f
natural *see* number
*nenrei hayamihyō* 81–4, 87
New Testament 47

New Year 106
Nichiren *see* Buddhism
night 99
Niigata (prefecture) 125
*nim* 147f
nine 54
ninety-nine 61f, 66, 102
Nishi 125
nomadism 118
*nombre marginal* 8, 22, 119f
North 64, 67
Noto (peninsular) 66
Nukisaki (shrine) 107
numbers: as abstraction 7; binary 10,
  35–8; cardinal 5, 7, 21–2, 109,
  closed system 36; complex 11;
  composite 10, 15; culture of
  46–62; irrational 11; Japanese
  (autochthonous 23–4, from
  Chinese 25–9); logic of 7; natural
  5, 14, 33 (*see also* infinite series);
  negative 11; odd 10; ontogenesis
  7; operational use 15; order
  (vicariant 7, 21–2, 99); ordinal
  5–6, 21–2, 99, 109; (connotation
  7; denotation 7); prime 161;
  symbols 1; words 15, 34; (in
  English 15); written 14; *see also*
  fractions; integers; language;
  names; *nombre marginal*;
  numerals
numerals 14; *kanji* 25–9; (cyclic
  system 27; frame system 27;
  *kun*-reading 27; *on*-reading 27);
  *see also* Arabic; English; French;
  place value; Yoruba

oak 50
*obi* 60
octave 6
*omikuji* 76–80, 88, 158
*on see* numerals: *kanji*
one-to-one correspondence 37
one 48–50; deletion 16, 25
Ontake 116
*onyōdō* 34–8, 43; *see also yin–yang*
oracle *see ekikyō*; *I Ching*
*ōrai* 35
ordinal *see* numbers
Osaka 49, 64, 115, 135

ounce 61
*oyakata see sumō*
oyster-beds 125
*ozeki see sumō*

*pachinko* 158f
paper *see janken*
*pars pro toto* 58
*pasocon* 158
passive 51
pawnshop 54
Pearl Harbor 95
peasants 67
perfection 55
philosophy: western 33
piano lessons 137
pilgrimage 116f
place value 17, 34
planets 112; *see also* five elements
Plato 14
pleasure 52
plover 56
pollution *see imiake*
pomegranate 49
poppy 49
post office 134
post station 115
pound 61
prevention 61
proverbs 49f
puzzles 12
Pythagoras 157

quantum mechanics 158
quartz (watch) 112

racing *see* car, horse
*rangaku* 12
ranking 150–4; *see also dan*; *kyū*
red 59
*remmei* 153
*renshi* 151
rice 3, 62, 65f; cultivation 4, 158;
  fields 125
right 125
Rin'ami (House) 122
*Rishhō Kōseikai* 73
risk 13
river 35
Roppongi 65
*ryō* 60

*saikoku* 117
*Saikyōto* 64
sake 61
Sakyō 125
*samurai* 67, 113, 157, 160
*sanjō sūji* 132
*Sanjūsangendō* 9–11, 102, 117, 119–21
*sanjūsankasho* 117
*sansai* 4
*sansukumiken* 143f
Sapporo 124
*sararīman* 116, 137
Saturn 39, 53
schools 150
scissors *see janken*
seal 119
seasons 52
Seiganto 117
*seijin no hi* 101
*seiki* 109
*seimeigaku* 3, 9, 68–73, 158–61
*seishikan see* death; life
*sekiwake see sumō*
*sen see* line
*setsubun* 101
seven 19–20, 47, 53f, 60; *see also shichi-fuku-jin; shichi-go-san*
Sevenoaks 65
seventy-seven 61f, 102
*shi* 41–3
*shichi-fuku-jin* 118
*shichi-go-san* 61, 101
Shigeko 68f
Shikoku 63f, 117f
Shimogyō 125
*shinbutsu shūgō* 104
Shingon (Buddhism) 117
*shinkansen* 115f
*shinra-banshō* 35
Shinto 61, 80, 104, 107; and Buddhism 104f, 157
ship 61
*shirei* 100
*shisoku see* addition; division; multiplication; subtraction
Shitamachi 124
*shōbunashi* 143
*shodō see* calligraphy
*shōgatsu see* New Year

*shōgi* 12, 140, 149f, 159
Shogun 99
*shonanuka* 100, 104
*shōnen* 72
shops 134
*shōrō* 100; *see also* Bon
Shōwa 110f
shrines *see omikuji*
*shū* (relating to time) 98
*shuzan* 20, 126f, 135
six 53, 60
sleep 89
sneeze 61
*sōkoku* 39–41
solstice 5
song (*uta*) 34, 43–5
*sora* 114
*sōshiki* (funeral) *see* death
*soshin see* ancestor
*sōshō* 39–41
*soto* 116
South 64
space 114–25
square roots 134
stable 105f; *see also sumō*
stamp 18–19
static 114
stitch 50
stone *see janken*
storm 4, 100
subtraction 17, 33, abacus 125f; *see also kagen*
suffering 52
*sūgaku* 7
*sūjikankaku* 136
Sumeru (Mount) 120, 125
summer 49
*sumō* 67, 151–3, 159; *banzuke* 152; *iemoto* 153; *heya* 153; *jūryō* 151; *kanroku* 152f; *komusubi* 153; *maegashira* 15, 153; *maku no uchi* 151; *oyakata* 153; *ōzeki* 153; *sekiwake* 153; stable 153
sun 4, 53, 99, 101
Suntory 161
swallow 49
sword-play *see kendō*
syllogism 51
symbols *see* number

Takeshi 67f
talisman *see mamori*
Tanaka 68f
*tanjō see* birth
*tanka* 43–5
*Tao* 36
tar 61
*tatami* 122
taxonomy 51
tea ceremony 136, 150
technology 3
telephone: directory 7; numbers 22
television 160
temperament *see kishō*
temples *see omikuji*
ten 54f, 60, 116
*ten'i* 72
ten thousand 50f, 55f
*tenchi* 35, 70
three 47, 60
thirty-three 9–11, 117f
thousand 50f, 56f
thousand and one 120f
thunder 37
tiger 52
time 4, 6, 28–9, 67, 96–113;
    arithmetical 112; non-repetitive
    96, 104, 113; repetitive 96, 98,
    100, 104, 107, 113; *see also*
    alternation; cycles; *kanji*; *ki*;
    measurement; *shū*
*tōhachiken* 144
Tōji (Temple) 122
Tōkaidō 114
Tokugawa *see* Shogun
Tokyo 64–6, 115, 159, 161
*tomurai-age see* death
topography 115
*toshigami* (year gods) 107
Toyama 49
transport 115
trigrams 36–7; *see also hakke*
*tsukebito see sumō*
Tsukumo (bay) 66
T'ung Shu (almanac) 112
Turkish 23

twenty-five 120f
twenty-four 60

*uchi* 116
*uchū* 35
*ujigami* 101
Ukyō 125
unstable 106
Uogashi 159
upper 125
*uranai* 76, 90
*uraya-san* 76

Vat 69, 161
Venus 39, 53
village 121
visiting card *see Meishi*

*wa-* 23
*waka* 43
*wakare* 72
*wasan* 8, 12, 23, 157
washing up 145
water 37, 39, 67
week 6, 98
weight *see* measurement
West 64; *see also* philosophy
wind 37
Wonderland *see* Alice
wood 39, 67, 121

*yakudoshi* (years of fate) 102f, 160
*Yama* 67, 116
Yama no te 124
Yamamoto 67
Yamato 23
year 4, 6, 112
yellow 39
*yin–yang* 10, 34–8, 51, 72f, 161
*yokozuna see sumō*
Yoruba 15
*yoshi* 70
youth hostel 145f
*yūkei* 124

Zen-Nihon 151
zero-suppression 17

Milton Keynes UK
Ingram Content Group UK Ltd.
UKHW031535071024
449327UK00005B/37